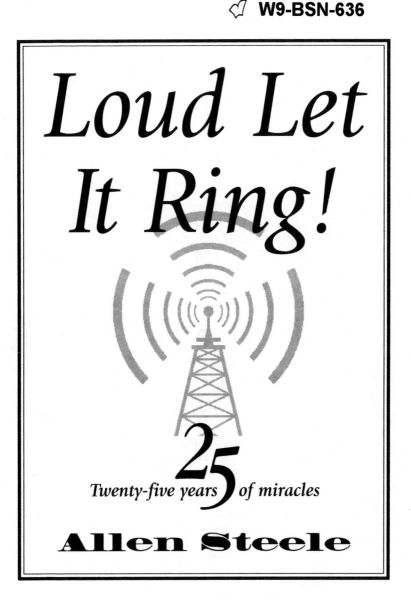

Loud Let It Ring!

Twenty-five *25* years of miracles

Allen Steele

Pacific Press Publishing Association
Boise, Idaho
Oshawa, Ontario, Canada

Edited by Kenneth R. Wade
Cover and inside designed by Michelle C. Petz
Typset in 12/14 Giovanni

Copyright © 1996 by
Pacific Press Publishing Association
Printed in the United States of America
All Rights Reserved

Library of Congress Cataloging-in-Publication Data:
Steele, Allen R., 1943–
    Loud let it ring: Adventist World Radio: twenty-five years of
miracles / Allen R. Steele.
    ISBN 0-8163-1335-0 (alk. paper)
    1. Adventist World Radio—History. I. Title.
BV656.S74   1996
269'.26—dc20   95-50945
                                                    CIP

ISBN 0-8163-1335-0

96 97 98 99 00 • 5 4 3 2 1

Table of Contents

A special note from

AWR

Sale of this book benefits the
AWR Program Opportunity Fund, which
helps increase worldwide radio coverage.

Thank you for your support.

Foreword

Adventist World Radio is the international radio voice of the Seventh-day Adventist Church. Over the past decade, it has grown remarkably to become a major international religious broadcaster. This growth began with the decision to build a shortwave station on Guam in the western Pacific Ocean.

No one is better equipped to tell the story of AWR than Allen Steele. His mixture of incident and fact recounts the story of AWR from its small beginning to its present considerable role in gospel broadcasting.

AWR has its origins in the conviction of the Seventh-day Adventist Church that the whole world should hear the "last-day message" of the everlasting gospel spoken of in the three angels' messages of Revelation 14. It is remarkable that this rather small but rapidly growing denomination has been able to conceive a worldwide radio outreach and then implement it.

The story of how this happened is worth telling, and this book does it in a fascinating way. Those who love to see the gospel going into all the world will also warm to its author and his personal account.

WALTER SCRAGG

On my desk are two large bolts and nuts that I use as conversation pieces and paperweights. These came from a supply that was used to fasten the antenna together at radio station KSDA on the island of Guam. At the inauguration of this facility, we had two one-hundred-kilowatt shortwave transmitters. Now we have four, giving KSDA access to at least 80 percent of the world's population. This book tells the colorful story of the phenomenal growth and development of Adventist World Radio. It also tells the personal story of Allen and Andrea Steele.

The story is filled with suspense, pathos, excitement, and irrefutable evidence of divine providence. Once you start reading, you'll find it hard to stop. You'll find descriptions of exotic places, frightening events, and intriguing cultures. But the real focus of this book is to illustrate the miraculous power of the gospel and the ministry of the Holy Spirit in the lives of men and women, young and old, regardless of where they live. The book will adequately silence those who have often made pessimistic and skeptical predictions about the usefulness of shortwave radio in reaching the world.

It is my prediction that this book will challenge and influence you to become a participant in the ongoing drama of radio evangelism. I am convinced that your faith in, and support for, the role of AWR as an indispensable agent in achieving Global Mission will be appreciably strengthened. The sooner you get started reading, however, the easier it will be for you to understand why I urge you to enjoy this book and discover the marvelous reality of God's intervention in human affairs.

My final comment regarding the story of AWR is best expressed in words I used at the time of the inauguration of KSDA— "A long-cherished dream is fulfilled, and many prayers are being answered." READ ON!

NEAL C. WILSON

Preface

This book is the result of a desire I have had for many years to share the many stories and events that have made AWR the great ministry that it is today. While I have tried to be true to the historical facts, I would like to emphasize that I wanted it to be my personal story because I believe the personal touch makes it more interesting. Therefore, I hope you will realize that many incidents relating to the facts of history are colored by my own experience and recollection and should be read with this in mind.

Such a story cannot be told without including the names of many people. In this case, most of the people mentioned are my dear colleagues from the AWR family. I thank them for helping to make the story possible. But one cannot possibly include the names of all those wonderful people who have had a part in making AWR history, so I hope that even though their names have not been mentioned, they still will rejoice in reading AWR's story.

I would especially thank my dear wife, Andrea, who was always a special part of our AWR team and who has permitted me to quote her so often throughout the book. She was also my on-the-spot editor and critic for this manuscript, making sure my recollections were correct. I also thank my sister and brother-in-law, Elsie and Buddy Blair, for their support through the years, especially for making it possible for me to pursue my college career and for being our main source of moral support "back home."

Neal Wilson and Walter Scragg are a large part of the AWR story, and I wish to thank them for their trust and support over the years. Finally, may I again thank the many AWR staff members who enter the story at God-appointed times. It was their appearance and participation that helped us to depend on Him to send the right people at the right time, which He always did. God bless you all!

Chapter 1

Could This Be the End?

"We've just been told a military coup is taking place." My colleague's voice was more distressed than I had ever heard it. "They are saying that the prime minister has gone to the National Guard barracks downtown to try to organize a counterattack. We should close the office and go home until the situation becomes clear." A sense of alarm overtook the four of us in the office as we gathered to discuss the news.

I ran to the phone to call my wife at home. "It must be a joke." She laughed at the mention of a coup. I assured her it was no joke, and she became serious. "Well, please come home right away. Be careful."

"Don't worry, dear. I'll hop the next streetcar and be home in a jiffy," I replied.

With one hand, I hung up the phone; with the other, I reached for my shortwave radio receiver. *If anyone can tell me what is really happening, it will be the good old BBC*, I told myself. I tuned in just in time for news at the top of the hour.

"Leading the news today is the latest on the apparent military coup in Portugal. For a direct report, we go now to our correspondent in Lisbon," said the lead announcer.

"Here in Lisbon, there is relative calm, but the military, under apparent leadership of General António de Spinola, has declared that the Caetano regime has been toppled and that the military is firmly in command, with the possible exception of some outlying barracks of the National Guard," reported the stringer from somewhere, maybe just a few blocks from me, in Lisbon. He went on to say that the key to the situation was whether all the military would bow to de Spinola's command or whether some would remain loyal to the old government.

All of a sudden, I realized that things could get really ugly. Aside from personal danger, what concerned me most was that Radio Trans Europe, the facility that broadcast our programs in Portugal, might be forced off the air. Was our mission to broadcast the Adventist message to Europe in danger?

My wife, Andrea, and I had been sent to Portugal three years earlier by the Seventh-day Adventist Church to begin the work of what was optimistically called Adventist World Radio (AWR). It would be many years before our project reached the majority of the world, but our broadcasts had already been attracting a wide audience all over Europe. Would the Lord permit this situation to silence His message?

Curious to see what was happening downtown, I raced to the balcony of our office. I gazed down seven stories at the broad Rua Braamcamp in front of our office building. It was very quiet, except for the comforting sound of clanging iron streetcar wheels ambling down the hill.

Then I spotted an armored tank at the corner. At the same moment, two fighter jets streaked overhead and off into the blue. Serious stuff!

Rushing back to the office, I grabbed my jacket and keys off the desk and headed for the door. Then I remembered the stack of mail on my desk. If I could just get this to the post office, maybe the postal service would be able to get it out before everything turned to chaos. It would be worth a try, anyhow, I decided, as I ran to my office and bundled up the letters and Bible lessons.

"The police may not let us go all the way to the end of the line," the conductor informed me as I jumped through the streetcar door. I noticed that there were only two other passengers.

"That's all right," I said. "I'll just go as far as I can."

The streets were empty of traffic. Usually a bustling, noisy city, Lisbon appeared to be on holiday. As we approached the central post office, there was more activity on the streets. This was closer to the heart of the city, and I felt better seeing people out on routine errands.

I jumped off the trolley and strode across the street to enter the post office rotunda. Just as I dropped the letters in the mailbox at the door, I heard *bang, bang . . . bang, bang, bang.* I hadn't heard that sound since army boot camp five years ago. Gunfire!

Immediately, as if a film had been turned to fast speed, everyone within sight started running. I did too. Around the back of the post office and down the narrow street I ran. Soon I was the only person on the street. Fortunately by then, I was within a couple blocks of the streetcar stop where I could catch a trolley home.

"This is the last *eléctrico* for today," the trolley conductor said nervously as I rushed breathlessly through the door.

"Great, I'm glad I caught you," I gasped. After a half dozen more people ran up to the car, it began its noisy, clanging trip through the cobblestone streets of ancient Alfama in the direction of home.

When I walked through the front door of our little apartment, I was nearly knocked over by Andrea, who bounded at me to squeeze me with a giant bear hug. "Oh, I've been so worried. Why did you take so long getting here?" she demanded.

When I told her about my trip to the post office, she

derided my explanation. "Don't you know it's dangerous out there? The radio keeps saying that everyone should stay home until danger is past."

"Oh, really," I said innocently. "Do you have any more news?"

"Come, our neighbors know all about it," she said as she led me to the back balcony. In Lisbon, the back balcony was the place to get all the news. Not only was it the place where laundry was hung out to dry; it was also where all the women hung out to share family news and neighborhood gossip.

Sure enough, our next-door neighbor was leaning out her window shouting bits of information to a woman on the next floor down.

"They say it should all be over by sundown," she said.

I interrupted the conversation to say, "I was just down at the central post office and heard some gunshots. Do you know what is going on down there?"

"Oh yes, that is just a couple blocks from the National Guard barracks, where the prime minister has been cornered by the army," she said. "You weren't really down there, were you?"

"I'm afraid I was," I said sheepishly. "But I ran for the last eléctrico up the hill, and I didn't see anything of the army."

"You were lucky. Who knows what might happen down there!" she scolded. Andrea held my arm tightly. "How are you doing for food?" our thoughtful neighbor asked as a quick afterthought. "If you need anything, let me know. We have things we can share." *Wonderful Portuguese hospitality in the midst of government collapse,* I thought. We assured her that we had plenty to keep us for a few days.

We spent the rest of the day inside the apartment reading books and preparing meals. Every so often, we would look out the front window or go to the back balcony to see if any more news had moved along the neighborhood grapevine. We included a session of prayer to ask the Lord to remember His work during this time of trouble and to protect Radio Trans Europe from political interference so that AWR programs could continue undisturbed.

Thursday, April 25, 1974, the day of the great military coup, was soon history. The uncertainty about how things would turn out continued through most of Friday, but by the weekend, things had settled down. It was a weekend of celebration. City residents formed spontaneous motorcades through the streets, horns blaring, to welcome the change. Others were in a state of shock. What would the future bring? Another dictator? Would de Spinola be true to his promise to inaugurate a democracy?

Andrea told me of a conversation she overheard on our street. "I've been a slave of this country all these years!" a man said. "I was afraid to speak out, even in my own family, because I never knew who might be an informer." Stories began to surface in the newspapers about blackmail, informers, and political prisoners during the dictatorship.

The country heaved a giant sigh of relief when national radio announced that the prime minister, the country's dictator, realized his days in office were over and surrendered to the Portuguese military. It became known as the Flower Revolution, because people in the city gave flowers, mostly red carnations, to the soldiers in the streets when they heard the news of the dictator's surrender.

"And listen to this," said Andrea as she read the special edition of the local newspaper, *Diário De Notícias*. "Only two people lost their lives during the two-day revolution, two soldiers who became victims during tensions in front of the central quarter." That must have happened at the very time you were at the central post office last Thursday, you naughty boy!" I winked at her and smiled. It was good to know my beautiful wife was concerned about me.

During the next months, we agonized with the people

of the turbulent country as they endured the vicissitudes of emergency, the harangues of political upstarts, and the uncertainty of a new political future. Immediately after the revolution, political parties of all kinds and philosophies surfaced. Communists came in force to push for their brand of politics. Socialist, Centrist, and Royalist parties appeared overnight. The beautiful monuments and façades were soon blanketed with political posters and graffiti.

The military junta, to its credit, was able to stay the course of the country until elections could be held. But the turmoil caused by the two-day Flower Revolution was to affect the country—and much of the world—for many years afterward.

But for the moment, we were glad we could get back to work preparing radio programs. Soon our days returned to a semblance of routine.

"We didn't lose even one minute of radio time through the whole episode," I telephoned church headquarters in Washington. Our ability to continue broadcasting straight through the revolution strengthened our resolve and our faith that God was blessing. It was a strong indication to us, among many that we were to witness throughout the history of AWR, that the Lord would jealously protect this special radio ministry so it could continue to search out His people in all corners of the earth.

Chapter 2

Play Radio

"Thumbalina, Thumbalina, tiny little thing,
Thumbalina dance, Thumbalina sing.
Thumbalina, Thumbalina, though you're very small,
When your heart is full of love, you're nine feet tall!"

My favorite Disney record came to an abrupt end, and it was time for me, the nine-year-old host of the Children Sing show, to warm up my audience for the next great tune about to come his way. I began to adjust the head of the electrical extension cord (held in place in front of my mouth by a wire coat hanger hung around my neck) that served as my microphone. I cued up the next record on the old wine red Airliner record player. But just then I saw my audience, my five-year-old brother, Lanny, head for the front door.

"Oh no, you don't," I threatened. "You can wait just a few more minutes till the end of the show! Now stay here." His face began to pucker into a wild yell for Mama, and I knew I had lost my audience. He taught me my first lesson in broadcasting: your audience has an attention span of only a few minutes. In broadcasting, we translate that fact into the acronym KISS: "Keep it short, stupid!" If your program is too long, the audience will turn you off, start doing something other than listening, or change the station.

"You used to bore your brother to death when you played radio," my mother would reminisce years later. "But you had radio in your blood."

"That blood had a difficult time circulating when you were born," she told me on another occasion. "So I asked God to save my little blue baby. I promised that if He did, I would dedicate him to His service."

It was a promise I had forgotten in the excitement of enrolling at Southern Missionary College in Tennessee. I decided my major would be business administration. Filled with awe and admiration for those administrators I had seen at the hospital where my mother worked as a nurse, I decided that the power and control, prestige and benevolence of a compassionate administrator, should be my goal.

Indeed, though I had no idea as a college freshman that it would happen, I eventually did end up as an administrator, suffered burnout, and pleaded for reassignment to bring relief from the pressures of dealing with people. But that would come years later. Right now, I had to surmount the challenges of a college career.

"You just aren't making it in accounting," my teacher said when he took me aside after class one day during my second year in college. "And your economics grade will be below what you need to maintain a place in this major.

"May I suggest you reevaluate your goals and see if you really want to continue in the business administration major?" he asked kindly. In my misery I didn't realize it, but God was using him to direct this young college sophomore into a path that he should have taken from the beginning. He did me a favor, for which I thank him to this day.

Later that night, in my distress, I begged God to show me the path ahead. "If I can't make it in this major, show me which one I should be in," I pleaded.

Miraculously, the answer seemed to unfold before me. God seemed to say, "Look at the courses you are taking.

Which ones do you like? Which are you doing well in?"
All of a sudden, it became clear. I had been doing well in
my communication classes: speech, oral communication,
radio announcing, debate. It was as clear as day. Why
hadn't I thought of it before? I headed for the campus
radio station.

"Hello, is anyone here?" I asked as I opened the door
to WSMC. It was Southern Missionary College's campus
radio station, and I felt uncomfortable because I was on
very unfamiliar ground. "Hello," I repeated.

Because it was still near the beginning of the term, the
station was not yet on its full schedule. I found myself in
a small, square office. A bare electric light bulb with a
round metal shade hung from an electrical cord that
dangled from the center of the ceiling. Two more steps,
and I was in the studio. Still no sign of anyone.

"Just a minute; I'll be right with you," a muffled voice
said. Then I looked through a large triple-pane window
and saw an authoritative-looking young man standing
before a large machine adjusting dials on its face. It was
the ten-watt FM transmitter that sent the station signal
out to the Collegedale valley.

"Hello, I'm Ed Mottschiedler," the station manager said
as he came into the room, extending his hand. "What can
I do for you?"

"I heard that you're looking for volunteer announcers,"
I said. "I'd like to try out." Either he saw the latent potential, or he was just hard up for announcers, because he
hired me on the spot. It was the beginning of my broadcasting career.

Some people called it a disease, a chronic behavior, a
mania, a cult. Whatever it was, it captured my imagination
then and always throughout the years that followed. "The
miracle of the ether waves," some old-timers called it.

Soon I became one of WSMC's regular announcers. This
was probably because I grooved on it so much that I was

willing to take announcer shifts at any hour—often talking other student announcers into letting me have their shifts. I had caught the bug. Within months, my dedication must have been obvious, because I was named director of public relations, then program director. I discovered that the broadcast industry has many directors. I also discovered that this was a misnomer. The directors, or chiefs, were very often also the braves who did the work. I was to hold the "director" title many times during my life. But whatever the title, in radio it was always hard work.

"Wow, what radio could do for the church," I said one day to Des Cummings, Jr., who had taken over as manager of WSMC. He was my mentor, and he knew exactly what I meant. We were kindred spirits.

"The church will need you in its plans to reach the world by radio," he said. We shared a vision that this electronic media could reach the hearts of millions of people for God. It was only a matter of selling the idea to church leaders. Surely there were some leaders up there at headquarters in Washington, D.C., who had the same vision. We found out later that there were more who didn't than did, and it would take a few convinced leaders and the enthusiasm of church members to silence the critics.

"We wish to ask you to be station manager next year," Dr. Gordon Hyde, chairman of the communication department, was saying. "We think you can handle the responsibility," he added. I was about to enter my senior year. I was giddy at the offer and grasped it without hesitation. That decision probably cost me extra time in school because it took me an extra half year to finish my degree. But I thrived on the responsibility and challenge.

I became convinced that the Lord wanted us to increase our coverage with WSMC. I soon started a campaign to increase the station's power so it could reach beyond our little college valley. Chattanooga, just across the mountain, waited for our message, I urged.

By the end of my senior year, we had raised enough money to buy a fifty thousand-watt Collins transmitter. We convinced Dr. Dewitt Bowen, an Adventist dentist, to permit us to raise a radio tower on some choice mountaintop property he owned near the school. From there, our signal would reach a wide radius that included Chattanooga. I will always remember Dr. Bowen as one of the many dedicated Christians who give generously and sacrificially to help the cause they love.

At the inauguration of the new station, the guest of honor and main speaker, US Representative Bill Brock, began his speech by saying, "I've been hearing about your plans for this new station for some time. I believe Allen Steele has become about as famous in Montgomery County as Bookie Turner [the infamous country sheriff at the time]!" It was a surprise statement that brought a round of laughter from the crowd. I was proud that the project had finally come to completion.

During my senior year, rumors began circulating in the Adventist Church that the church leadership was thinking about the power of radio to "preach the gospel to the world." It was exciting news. But could it possibly be true?

"There are indeed some leaders in our church who feel it is time we heed the call of Elder H. M. S. Richards, Sr., of the *Voice of Prophecy*." It was Elder C. E. Crawford, a church official from the Southern Union office in Atlanta, speaking at a school assembly. "The Bible says this gospel of the kingdom will be preached in all the world. The only way we can see that happening," he asserted, "is to utilize the latest technology to put this message into the ether waves."

It was a great speech, but I wanted to know *when*. I was about to graduate, and I needed to know now. "I don't know exactly when, my friend," he told me afterward. "But it is inevitable if we are going to reach the world."

"Right on," I replied.

No more news came about this idea, although Des and I discussed it quite often. My graduation day finally came, and I had nothing concrete to lean on as far as employment in Christian broadcasting was concerned.

With nothing else in sight, I traveled to Memphis to see if I could find a job to keep me busy and pay for advanced studies in communication. My plan was to study for a master's degree at Memphis State University.

I read that there was an opening for a TV cameraman at WKNO-TV, the university's television station. To my surprise, I was accepted and soon found myself on the staff as a kind of apprentice. My training would involve work in all departments of the station: production, graphics, public relations.

I was learning a lot and at least had a job until the next semester of school would begin at MSU. I soon discovered that TV moves much slower than radio. That is, there are many players in TV production, and each one has a rather boring routine until production time comes. Then everyone goes into action to bring the project to fruition.

Radio production was more creative, I decided, unless you were actually a producer of a TV show. The producers had all the fun, it seemed. The rest of us just waited around for a producer to come on the scene and decide that we would finally produce a show.

Alas, I had had only four months on the job at WKNO when I received a surprise letter that interrupted my television experience and my education.

"Greetings," it began. "You are hereby instructed to report to the United States Army Induction Center for your physical examination, from whence you will travel to Ft. Campbell, Kentucky, for entrance into the United States Army!"

As a Seventh-day Adventist, I was inducted into the army as a noncombatant. But after my basic training and four weeks of advanced individual training as a medical corpsman, the unit personnel officer discovered he had a TV cameraman on his hands and gave me orders to transfer to the Infantry Training School at Fort Benning, Georgia.

"Go get a mop and mop this studio floor," said Sergeant Brown. It was my first assignment in the "spit and polish" TV division at the training school. It became a recurring order, and I was quickly becoming disillusioned with television work. Occasionally we would take our cameras and mobile production studios out into the field, but I longed for the immediacy and intimacy of my radio microphone.

"Well, I don't know if you deserve it or not," Sergeant Brown said lackadaisically, "but I have a notice here that you are on levy for Europe!" I could hardly believe my ears! Just when I was at my lowest in the US Army, the Lord came along and arranged a free trip to Europe.

"I believe I deserve it, sergeant," I said politely and took the orders from his hand.

"Good luck to you," he said. Within two weeks, I was on a troop charter plane headed out over the Atlantic to Frankfurt, Germany.

My time in Germany started with a bang. About twelve of us soldiers were being escorted down the street from the Frankfurt clearing depot to the Hauptbahnhof Central Train Station, walking in single file, each carrying a huge duffle bag full of belongings over his shoulder. A big black Mercedes-Benz came up beside us, and a woman yelled out the window, "Hurrah for the U.S. Army!"

Well, it's nice to be wanted, I thought. But I was only allowed that one thought. The soldier in the lead was so distracted by the woman that he walked right into a telephone pole. It started a chain reaction, and we all stumbled and fell like dominoes. What a graceful way to enter Germany!

I was stationed in the town of Hanau, but I made the SDA Servicemen's Center in Frankfurt my headquarters on weekends. From there, it was only a day's drive in my blue VW "bug" to most of Europe. Before my term was up nearly two years later, I had visited every country in western Europe except Finland and Portugal. It was a good background for what was to come later and prepared me well for a new assignment in the Lord's army: the Seventh-day Adventist worldwide workforce.

Chapter 3

Two Loves

"Well, where do you want to land in the USA?" asked the army travel clerk. It was the one time in my life when I was completely free to decide my next direction in life. After thinking over my options, I decided that if I really wanted to pursue my dream of working in Adventist radio, I should see what I could find in Washington, D.C., home of church world headquarters.

It seemed it would also be an ideal place for finding good Christian friends. I knew the city had many Adventist churches, an Adventist college, and several Adventist hospitals. All this in addition to the excitement of the nation's capital city.

"Washington, D.C.," I answered, with only a little hesitation.

I arrived there with no friends, no contacts, no home. But soon I had them all. My goal was to find a job in Adventist radio, if possible, and to get involved in church youth activities. Once again, I was forced to rely totally on the Lord to lead. He came through, as usual, on all counts.

Fortunately, I had my discharge allowance from the army, so I had funds to survive for several months. I found an apartment near Columbia Union College, started checking out the churches in the area, and started my job search.

Radio jobs were rare, so I applied for a job at Leland Memorial Hospital and was accepted as an assistant in the personnel department. My main duties were to interview job applicants and keep the personnel records in order.

"Would you be interested in leading out in our public-relations projects?" asked Leonard Coy, the hospital administrator, a few months after I began work. Soon I was arranging health displays, holding stop-smoking clinics, and editing a health journal for the Leland Medical Group Foundation.

I knew I couldn't give up my love for radio work, so I sought ways to make contact with church headquarters to see if I could even volunteer to make radio programs. I had friends at several college radio stations who had organized the Adventist Radio Network (ARN).

"What if we could have a weekly press conference?" I asked Don Dick, the leader of the ARN group. "We might even be able to get a weekly program by the president," I added to sweeten the offer.

"Let's go for it," came back the reply. "I think the stations would be very interested in it." As usual, I was full of ideas, but now I had to come through with something concrete. Elder Robert Pierson, the General Conference president, was a father figure and very interested in communicating with the church.

"I'm intrigued by the concept of having regular communication with the church," he said when I approached him. "My main problem is to find the time to do the recording." I realized that he must be one of the busiest people in Washington.

Five days later, he called me back. "I've been thinking it over, and I definitely want to do this. Only one thing. The only time I can do it is first thing in the morning, before office hours. Is that possible?"

"Of course," I said, wondering how I would manage putting in a recording session before I even started work for the day. Where there's a will, there's a way, and soon we were in the routine of meeting once a week at 6:00 a.m. for the recording sessions of *Let's Talk It Over*. I considered that morning rendezvous with the president as my morning devotional.

One day I was sitting in the little windowless room that served as our recording studio in the South Building of the General Conference office complex. It was furnished with an institutional gray table, two gray chairs, and a small portable recorder on which I did the recording and editing. I was concentrating on editing the tape by marking, cutting, then splicing the recorded magnetic tape. Suddenly, I felt a pull on the tape in my hand, and the pulling continued. Something was eating *Let's Talk It Over!* Then I got down on my knees and watched the tape as it was pulled steadily away—up into the wall heater. The heater fan had caught the end of the tape in its draft and was slowly winding it around the fan shaft! A whole program had to be rerecorded, I sheepishly informed Dr. Pierson the next day.

My second radio project was a press conference. I discovered that there were several young people who would love to have the opportunity to grill church leaders on hot topics facing the church. Soon my team of Rosie Bradley (from the official church paper, *The Review and Herald*), Pat Horning and Chuck Scriven (from the new youth magazine, *Insight*), and Kit Watts (from the Sligo Church staff) were regularly interrogating church leaders at the weekly Adventist press conference.

I was back in love with radio and enjoying the fellowship of our vibrant church community in the capital. Sensing a need to serve the church even more, I decided to try a new idea out on some of the young people I had met.

"Sounds like a great idea; count me in," was the response I kept getting. It appeared to be an idea whose time had come. The plan was to form a group of senior young people into a traveling troupe that would visit smaller churches in the area on Saturdays to provide the Bible-study hour and church services.

At the organizational meeting, I marveled at the talent we had pulled together. Here were musicians, teachers, editors, medical specialists, media specialists, aspiring lawyers, and other professionals. We named our group The Wider Circle, taken from a poem or book one of the group had recently read. In no time, our church-service strategy was clear: different members of the group would assume the personalities of various Bible characters with whom Jesus dealt in the New Testament. The personal experiences of Mary and Martha, the rich young ruler, Paul, and blind Bartimaeus became the meat of our message. It was a great success, and the group continued even after I left the area two years later.

One night we had a Wider Circle practice session, and into my life came a new young woman. Andrea Grover wore her hair up in a bun, with a curl hanging over each ear. She played guitar and was willing to sing. I remember welcoming her to the group and thinking nothing more about the meeting.

Until eight days later. While sitting in my office at Leland Memorial, I picked up the newspaper from the Washington Adventist Hospital. I noticed on the back page a picture of the new secretary to the public relations director. Lightning struck. It was the same young woman who had appeared for the first time at our Wider Circle rehearsal that week. She was beautiful! Wow, what was I waiting for? I dropped everything and drove over to the big hospital in Takoma Park.

"I didn't see you at the Wider Circle meeting this week and was wondering what you think of our group." It was the best opener I could think of.

"I think it has a lot of potential," she said, giving me the impression that she thought it was great. Only years later did she confess that she really wasn't so sure about the group. She said that she probably would not have returned if I had not invited her.

It was soon clear that I had found my second love. The first was radio; the second was this charming, lively young woman. Soon she took first place, radio second. I had no idea at the time that my two loves would become as compatible as they have, but through the years we've discovered that the Lord had a special radio work for us both to do.

"It looks like the church may finally be able to use radio to reach all of Europe," James Aitken, secretary of the church's radio-TV department, said excitedly as he leaned over his office desk. "An old friend of the church, Mr. Tremoulet, has been able to get permission from the Portuguese government to install a shortwave transmitter, and he wants to sell air time to us. He appreciated the help our church gave him and his family after World War II, and now he wants to help us by putting our programs on the air over his own Radio Trans Europe. If we can move fast enough, we will have our choice of air time."

"Praise the Lord, this is wonderful news," I said, trying to sound as supportive as possible without revealing my excitement about the possibility of my own involvement. "If you need someone to go over there, I'm certainly available."

"That's precisely why I called you here today, Allen." He looked me straight in the eye. "We're wondering if you would be willing to go get these programs started over in Lisbon? Your experience and exposure to Europe uniquely qualify you for this job. Would you be interested?"

It was the invitation I had been waiting for, praying for. "Indeed I would." I was afraid it would not appear professional to exhibit too much enthusiasm. "When do we start?"

He looked at some papers on his desk and thought out loud: "Well, let's see. The station has promised us our choice of air time if we are the first to sign up as partners in this project. So I hope we can sign the contract by early next year or, at the latest, by spring. It is now December, so we would most likely need you over there by summer. Yes, let's start working on a summer start-up."

Much to his consternation, we were not the first to sign up—another religious organization from Scandinavia beat us to it by several months and would always have first choice for the best air hours. At any rate, Mr. Tremoulet's schedule was far more advanced than ours, and I knew I would have to work fast to meet a summer deadline.

I had already invited Andrea to come meet my family in Florida over the Christmas holidays. I knew now that it would be more than just a visit.

"Will you marry me?" I spoke those words on a moonlit night under the stars in my father's little orange grove in Ft. Ogden, Florida. He named that stand of trees "Honeymoon Grove" in our honor. Our marriage was set for the last Sunday in May 1971.

The new year was a glorious one. In spring, cherry blossoms were like mountains of lace around the reflecting pool in downtown Washington, and the Maryland suburbs were graced by mounds of azaleas. Our wedding day was only a few weeks before our planned departure for Europe.

We included our Wider Circle friends in the wedding service as musicians and readers of appropriate Bible texts. To avoid the usual wedding pranks that accompany American weddings, such as "Just Married" signs painted on the groom's car and the tying of old shoes and cans to the rear bumper, I parked my car in Rock Creek Park near a footbridge opposite Capital Memorial Church. After the reception, my brother took us the few blocks in his car to the little bridge, where we crossed over by foot and jumped into my car for a clean getaway.

During the spring and summer, we looked for books about Portugal. We tried to learn everything we could about the great adventure we were about to begin for the church. We discovered that many Americans did not even know where Portugal was. Usually they thought it was in Central America. "Portugal—isn't that near Panama?"

A debate had sprung up about what the new radio activity in Europe should be called. Some thought it should be *Voice of Prophecy* or *Voice of Hope*. Others thought we should be more straightforward and have a name with *Adventist* in it. Working through the decision-making process was Walter Scragg, assistant secretary of the radio-TV department. We found him a fascinating Australian who possessed a most delicate sensitivity about the meaning of words. He made me feel important when he asked me what I thought about the name we should use. "Let's call it Adventist World Radio," he said. "It has a certain ring, and I think it will wear well."

Walter became our mentor in the radio ministry, our main supporter at church headquarters, and later, a major force behind AWR's most exciting years. We immediately appreciated his ability to make us feel an important part of the radio team.

"Slippers in the salad bowl—that should work," I was quoted as saying in an article in *Insight* magazine. Pat and Kit, our Wider Circle friends, had come to record our frantic packing for the trip to Europe. While Pat was busy digging out all the information she could about the radio project, Kit was trying every angle for photos. She was on the bed, on top of the dresser, and in our faces. They watched us pack dozens of wedding gifts in a big box that would be put into storage. We didn't know it at the time, but it would be five years before we would have the opportunity of using those gifts.

Within weeks, *Insight* published their article: "Radio Project: Allen and Andrea Leave for Lisbon." In the same issue was a news brief about Mary Ellen Irwin, wife of Apollo 15 astronaut Jim Irwin. She had missed seeing part of her husband's moon walk because she was busy teaching a Bible-study class in the Nassau Bay, Texas, Seventh-day Adventist Church. Commenting on the fact that she wasn't able to see his first steps on the moon, she said, "Jim is committed to his mission, and I am committed to mine."

That's what life is about, I thought. *Being committed to a mission*. We were leaving a very comfortable, enjoyable lifestyle for a pioneering mission, and we had no idea what was ahead of us, what loneliness we would feel, how we would survive in another culture.

As we headed for the airport, I was amazed to see Andrea carrying a large Tupperware container filled with books. "We forgot to mail these with the other stuff," she said sheepishly, "and I know we're going to need them."

Departures were much harder for Andrea than for me. She had lost her real father in an automobile accident when she was nine years old, and her mother had died seven years before I met her. She had found confidence and permanence in Washington by "adopting" the Torsten Lundstrom family as her own. Now she was leaving her best friends and new family to go on a faraway mission with a man she hardly knew.

"I think being a missionary is not so much the difficulty of going to a foreign land," she confided as she slipped her hand into mine on board the airplane. "The hard thing about being a missionary is being separated from family and friends." Her tears emphasized the fact.

Chapter 4

On the Air

"What a beautiful city you have," I said to the union president as we looked over part of Lisbon from one of its seven hills. We were thoroughly entranced by this colorful city that sits on the estuary where the Tagus River empties into the Atlantic. Lisbon quickly became my favorite city in Europe, and it remains at the top of my list today.

"Houses or apartments for rent are easy to spot," the president told us as we started searching for a place to live. "Owners place little square pieces of paper in the middle of every window."

He was taking us to the Belem (Bethlehem) section of town to see some apartments that were advertised in the newspaper. When we stepped out of the car, we were immediately surrounded by four ragtag children begging for money. One little boy of about three was wearing only a T-shirt; a little girl about the same age was wearing only some underpants. All four were dirty and unkept.

"Don't give them anything," the building's porteira, or landlady, called out the window. "If you do, they will never stop pestering you," she added as she ushered us into her building. In time we found out that porteiras are very important people in Portuguese life. They manage the building and live on the ground floor to keep an eye on everything. If you tipped the porteira well at Christmastime, you could be sure that everything you owned would be kept under protective surveillance.

We decided to take an apartment with a beautiful view of the river and the famous Belem Tower, which in ancient times served as the tollhouse for ships entering and leaving Lisbon harbor. Just two blocks away were two more equally important landmarks of Lisbon, Geronimos Monastery and the Monument to the Discoveries. The huge monastery is the most famous in Lisbon, and the monument has statues of all Portugal's famous explorers, headed by Prince Henry the Navigator, lined up on the prow of a ship.

The one-bedroom apartment was the first of three that we inhabited in the five years we lived in Lisbon. After we had lived in Belem for about a year, a bumper-car concession set up in a vacant lot behind our apartment building. It soon became the bane of our existence. It opened for business around midday, then continued until two in the morning. The kids bumping into each other were noisy enough, but worse was the constant blaring of American rock music. The operator must have especially liked "O Mammy, Mammy Blue," because its tune assaulted our ears every night until the wee hours of morning.

"I can't take it any longer," I announced to Andrea. "We've got to move if I am going to keep my sanity." So we started searching. We were enchanted by Alfama, the oldest part of the city, near the hilltop Castelo de São Jorge (the Castle of St. George), the legendary first site settled in the city. It had narrow, winding cobblestone sidewalks, stairs, and streets and was the subject of most of the picture postcards sold in Lisbon. Clothes were hung out on clotheslines moved by pulleys between buildings, all of which stretched over sidewalks where pedestrians passed below, ever subject to drips of water falling from the clothes put out to dry.

It was here one day that we noticed an apartment for rent, right on the streetcar, or eléctrico, line on Rua Voz do Operário (Voice of the Workers Street) and across from the São Vincente Church. What fun, we thought, to be in the ancient Alfama quarter. Here we could get a taste of old Lisbon.

Unfortunately, we didn't visit the place on a Saturday, because we didn't discover until our first weekend there that our new neighborhood was the noisiest place in the city on Saturdays. The huge Lisbon Flea Market ended, or started, depending on which way you were going, right in front of our house! Here one merchant parked his truck every week, set up his microphone and loudspeakers, and hawked woolen blankets all day Saturday.

Our one day of rest became a headache. For this and other reasons, we decided to look for our third home.

This time we headed for the newer part of the city, Benfica, renowned in Europe because of its championship football team. Here we found a place overlooking the Monsanto city park, near a Metro (subway) stop and only four blocks from the Lisbon Zoo. I liked to tell my friends about waking in the middle of the night to the sound of a lion's roar. This became our most enjoyable home, and we were very comfortable there until we left Portugal.

But our first home was our Belem apartment, where we settled in quickly and turned our attention to the work of AWR. As part of the agreement with Radio Trans Europe, we were allocated an office in a suite occupied by the station owner.

Our office was on the top floor of a residential building that was built about 1900. It was on busy Rua Braancamp, which was one of seven spoke roads leading out from the statue of Marqués de Pombal at the foot of the Edward VII Park in the newer upper section of the city. Pombal was prime minister of Portugal when the disastrous earthquake of 1789 hit the city. He was able to galvanize the shocked people into rebuilding the city, and he was one of the most venerated statesmen of Portugal.

Some of the residences in our building were converted to office suites, similar to ours. Others were still occupied by old families of Lisbon who seemed ensconced for the duration. The country had a law that tenants could not be evicted from their homes under any circumstance, nor could rent be increased, so all over the city people lived in their apartments until they died, when building owners would have their first opportunity to utilize the space in another manner.

To arrive at our office suite, one had to either ride what my wife called the "proper Victorian English lift" or climb seven stories of stairs. Since the lift was more like a giant cage, with ornamental iron bars and beveled windows, one could easily throw words of greeting to people waiting at each floor. Or sometimes even be the first to greet residents as they opened their door to leave home for the day. Often I raced the lift by running up the stairs, circling my wife in the cage, reaching our office floor well before the elevator.

"Look at this office." My wife was awed when we first entered our assigned place at the end of a long hallway. "Aren't we fortunate to have the kitchen and pantry as our headquarters?" she said matter-of-factly. We poked into the corners of a room that must have once seen the preparation of beef, pork, rabbit, chicken, and other meals. A faint smell of olive oil permeated the place.

"Look at this horrible red carpet," she went on. What caught my attention was the single electric cord hanging in the middle of the room with a round metal shade over a single light bulb. It reminded me of the first time I visited WSMC nearly ten years before.

The station owner rarely came to Lisbon. He left things in charge of Madam Branca, a very capable, no-nonsense administrative secretary. We soon learned that no one could put anything over on Madam Branca, and we did our best to please her in all things.

She became a delightful part of our life, being the person who would care for many of the business matters. She always gave us good advice and usually was the final word on any important subject that would come up. She was also a very good guardian of the office door. We could depend on her to handle anyone who came in.

One day, however, I was the only person in the office. I heard the doorbell ring and ring again. Finally, I decided to find out who was there.

A family with three golden-haired children awaited me. They explained that they were from Sweden and were looking for the director of programs for another religious organization. I told them that he wasn't there, but that he might show up later, and perhaps they could just wait for him in his office. After showing them to his office, I promptly forgot about them. . . .

Until the next day, when Madam Branca came storming into our office. "Did you invite those people into the office?" she demanded.

"What people?" I replied, desperately trying to figure out whom she was talking about. Then all of a sudden I remembered the Swedish family.

"Do you realize they spent the whole night in that office?" she retorted. "You must never let people come into any office unless our staff are there," she instructed and turned heel to leave in a huff.

In addition to the office, we had access to the production studio across the hall. My first priority was to start collecting programs to put on the air. Our studios in Paris, France, and Darmstadt, Germany, were assigned the bulk of the program production. In Paris, Roger Fasnacht and Bernard Pichot were busy preparing programs in French and a half dozen other languages. Bernard became a good friend and colleague and was still at his post making programs when we came to serve in Europe a second time; twenty-one years later.

The programs were sent to us by mail or hand delivered, and then it was my duty to time and edit them, add the AWR station identification, and package and deliver them to the broadcast studio in Sesimbra, about twenty miles south of Lisbon.

I got right to work, because our first day on the air was set for October 1. By that date I had to have at least the first week of programs ready. Roger and Bernard were coming down from Paris to be with me the first day on the air, and when they arrived, they planned to bring several boxes of recorded programs for the next few months.

We were obliged to follow a delivery system that the other religious group had devised for getting programs to the broadcast studio. They would combine their taped programs into one-hour blocks, put them on a fourteen-inch metal reel, and place them in a cardboard file box that fit snugly into white plastic shopping bags. To save confusion on the part of the studio technicians, we decided to use yellow shopping bags from a supermarket to identify and deliver our programs to the studio in Sesimbra.

"Someday you ought to write a book about AWR and call it 'A Thousand Yellow Plastic Bags,' dear," Andrea said one day. I didn't think much of that idea, preferring the more exotic sound of "Towers on the Tagus" or "Antennas in Alentejo."

All our activities hurtled toward that October 1 beginning date, which arrived almost before we knew it. I worked on programs most of the night of September 30 and finally was ready to dash down to Sesimbra and deliver the programs an hour before the first was to be broadcast.

The studio technicians quickly took the programs from me as I entered the studio, and they ran to put the first program on. That first program was in Italian, and I recorded that in my diary. Years later, when some of us tried to set the historical record straight, it was suggested that the first program might have been in Russian. An article appearing in the *Review and Herald* said the first program was Ukrainian. But that first night on the air I wrote in my diary that it was Italian, and I trust my notes more than any other source. I think it is very appropriate that the first language was Italian, given the significant role that country was to play in the story of AWR.

We thrilled to hear the identification tune AWR had chosen, "Lift Up the Trumpet and Loud Let It Ring," as it first pierced the airwaves. For the first time, the Adventist message was in the air over a large portion of Europe with the trumpet theme, borrowed from the *Voice of Prophecy*. Some years before, Adventist programs had been aired on Radio Luxembourg, but this was the first full-scale attempt at reaching all of Europe.

It was a high day, and I was exhausted. We made our way back into Lisbon over the Ponte Salazar, the Golden Gate bridge of Portugal.

"Well, what happened—are we on the air?" James Aitken was on the line the next morning anxiously wanting to know if all had gone according to plan. I suddenly realized that in my excitement and exhaustion I had forgotten to let Washington know that the historic moment was a success.

"Yes, sir," I said. "Everything went well, and Adventist World Radio is on the air." Now all we could do was wait to see if there would be any response.

Chapter 5

A Bold New Move

In 1971 Robert Pierson joined Neal Wilson, president of the North American Division of Seventh-day Adventists, in writing an article to launch AWR's first project. Their article, titled "A Bold New Move to Finish the Work," appeared on the front page of the *Review and Herald*.

The project "marks the largest single endeavor the church has ever made in international broadcasting," they wrote. "For the first time in recent history the General Conference is launching out on a large undertaking that is not fully funded." They pointed out that the church would need donations of $128,000 to care for the first year of thirty-two broadcasts per week in twenty languages. They approached the theme positively by saying, "For several years many of our committed laymen have been challenging church leaders to think big and undertake a bold, faith-filled step forward toward finishing the work." They had been encouraged to start such a project, they asserted, by church members who said, "There are hundreds of thousands of dollars in Adventist pockets and bank accounts that will be forthcoming when we are confronted with such a challenge!"

Even with the top administrators behind the project, the future of AWR was far from certain. Andrea and I were asked to go to Lisbon for only one year, with the understanding that if enough money came in, we would be asked to stay another year. We were well aware that shortwave radio was an unproven resource in the church. Many leaders, especially those from North America who had no idea what shortwave radio was, openly called it a waste of money, time, and energy.

Presidents Pierson and Wilson had accepted the challenge, and now we were in Lisbon to carry out the mechanics of the project. After the excitement of the first day, we continued preparing programs and waiting for letters from listeners.

Soon the letters began coming in. Before long, we found that the highpoint of the day was the arrival of the mail carrier.

"Listen to this," I said excitedly to Andrea. "This man says he listens to us every night from the tallest building in Europe—the tower of the Cologne [Koln] Cathedral!" Employed as a night watchman at the cathedral, his letter told how each night, when he climbed the tall bell tower on his rounds, he took his portable shortwave receiver and listened to the world, a world that now included AWR!

Another person whom we learned to appreciate was Arthur Cushen of New Zealand. He always had his ear tuned to the world. In spite of (or perhaps because of) his blindness, he is one of the world's most renowned international radio monitors. He was quick to let us know of his reception of our broadcasts from Europe and years later was one of the first to report hearing AWR's station on Guam.

Many of the first letters were from a special group of people called DXers. DX, in radio lingo, stands for "distance unknown." These people are avid radio listeners who write to stations they've heard, describing programs they've listened to. The station then sends them a QSL, or verification, card. DXers collect these cards as symbols of their prowess at their hobby.

Some church broadcasters don't consider these people as serious listeners. Why should we program to hobbyists, anyway, is the reasoning. In fact, we know of a number of DXers who have become Adventists because they were first attracted to AWR by a DX program.

In Austria, Heinz Haring became a regular AWR listener shortly after he took up DXing. He requested the Bible course and started studying with his wife, who also had an interest in listening to international radio. It wasn't long before this family decided they should identify with the Seventh-day Adventists. They continued to be official monitors for AWR for many years and were so supportive that they often spent their yearly holidays in visits to AWR stations and offices.

It didn't take us long to realize that DXers could help us evaluate our signal propagation—that is, how well our broadcasts were being received in their intended areas. Most other major broadcasters already included special programs for DXers in their schedules, so we decided we should have one too. But how could we produce one with our meager resources?

"I'll just write a letter to several DX clubs and ask if they would be interested in cooperating with AWR to produce a DX program for Europe," I said to Andrea confidently. "The first one that responds will get the air time on AWR." Actually I didn't feel very confident that we'd get a response, since we were so new and our content was religious.

But just a few weeks later, Andrea brought the mail in and announced, "Here's a letter from Clive Jenkins, president of the World DX Club in England. He says his club would like to cooperate with us on the DX program." It was the beginning of a relationship that lasted quite a few years, and the resultant World DX News on AWR became one of the popular DX programs in Europe.

Years later, Adrian Peterson developed another extremely popular DX program called WAVESCAN. Adrian developed his own style of program when he was in charge of radio programs for AWR-Asia out of Poona (Pune), India. His first program was called RADIO MONITOR'S INTERNATIONAL, and it gained some notoriety in Asia and, indeed, other parts of the world where people could hear it.

In the 1990s, Adrian developed WAVESCAN into a leading worldwide DX program. His knack for including interesting historical background information about DX stations and personalities made his program a special catch for radio listeners.

Just as in the DX world, we were pioneering for AWR in the whole field of international broadcasting. We had to accept that many aspects of our work might not come to fruition until years later. After all, to "broadcast" means to sow seed, in its first English definition. We just felt fortunate that we had this station in Portugal to sow seed at a time when European governments exercised tight control of the airwaves. This was the first time (except for a few years earlier, when Victor Cooper was able to get some programs on Radio Luxembourg) that the church had opportunity to obtain large blocks of air time in Europe.

One of the best examples of this concerned our programs in Greek. The early radio programming from Portugal held out much promise for Greece, where Nick Germanis, a transplanted American, was anxious to make a difference in that Orthodox country. Adventist church membership there numbered less than two hundred. He was anxious to do everything possible to get Greek programs into production.

"Fortunately, there is a Greek program by the Greek government on AWR," I told him by telephone. "We can put your program on the air immediately after the government-sponsored newscast."

It seemed like a good strategy. For over five years, we were able to take advantage of the Greek government's program to establish an audience for ours. In those days, Greece was ruled by a dictator, and their programs from

Portugal were an attempt to favorably influence Greeks abroad about the government back home. A small team of Greek broadcasters had offices with us at the offices on Rua Braancamp.

Response was good. Letters from exiled Greeks in Austria, East Germany, Poland, and other countries came in regularly. They seemed thirsty to hear their own language—they were exiles from home who were open to new ideas.

"Listen to this man's story," Andrea said to me one day. "He says he finally got the Bible we sent in with tourists. But it was the end of a long saga. He was so inspired by the 'book' we keep talking about on the radio that he began to search for one in Leipzig, the city where he lives." Her eyes danced with animation as she continued the story.

"You remember, he's the man who asked for a Bible, but when we sent it, it was returned by the East German post office with 'addressee unknown'?" She continued. "Well, he wanted a copy of the Bible so much that he searched the bookstores in the city. Then he started asking at churches. He found an old Bible in one church, but it was in old Greek, which he couldn't read. Now he says he is a new man because of the Bible we sent."

Later we found out the man had been baptized and had become a member of the Leipzig church. Similar stories filtered out of eastern Europe during our tenure in Lisbon.

The greatest disappointment to our church leaders in Greece was that no letters came from Greece itself. We keenly felt the disappointment ourselves, and when it came time to leave Portugal, we carried the disappointment with us.

Then in 1989 we were told the most wonderful news. On our way to a meeting in Africa, we decided to stop over in Europe. We put Athens on our itinerary and enjoyed visiting the ancient ruins at the Acropolis and around the city. On Saturday, we decided to go out and search for our church.

We only knew the name of the street, which we remembered from hearing it so many times on the radio programs. We soon found Keramikou Street and decided to walk several blocks from one end to the other until we saw the church. Andrea took one side, and I took the other.

After two blocks, I noticed a man carrying a Bible, so I motioned to Andrea. She joined me as we followed him to the Adventist church. Inside, we received a warm welcome and were asked to give a report about AWR. After church, an elderly gentlemen came up to me.

"I'm Christoforides, the one who used to make programs for Portugal. Do you remember our contacts by telephone?" Indeed I did. With a rush, the disappointment of all those years of no letters came back to me. "My wife and I invite you to our house for lunch; can you come?" he asked.

We ate well of the vegetarian Greek dishes set before us. After lunch, we talked about the church in Greece, the difficulties of church growth in an Orthodox land, and their plans for their continuing years of retirement. Then he had something special to tell me.

"You remember our great disappointment that the radio programs did so little for Greece?" he asked. "Well, we should have had more faith in the Lord. You know what happened?" By the twinkle in his eye, I could tell he had exciting news.

"When the repressive government was thrown out and democracy returned to Greece, its native sons began to return. Some of those returnees were by then Seventh-day Adventists. They came back to their families in Macedonia. They shared their new beliefs, and that is why today we have three small congregations where before we had none!" His eyes danced, and his cheeks creased. "All of a sudden, our number of churches doubled, and we had a new area of the country open to the gospel and churches planted by God Himself."

Tears came to our eyes as we rejoiced together.

Chapter 6

A Vida Portuguesa

"Here, kitty, kitty. Here, kitty, kitty." Andrea can never resist a quick friendship with any cat that is within view, and this one, a Siamese with a crooked tail, was especially fetching. "Here, kitty, kitty," she said again, but too late because the cat had slipped through the ornate gate that entered a drive where a black Toyota sedan was parked.

We were on one of our occasional walks through the Belem neighborhood near our apartment building. The sloping hillside was bathed in afternoon sun. Large- and medium-sized houses, most of them with whitewashed walls and red tile roofs, graced the slope overlooking the Tagus River estuary. It was the favorite neighborhood for embassy staff housing and well-to-do Portuguese. Some houses served as both residence and embassy for small countries.

"Well, hello, how do you do?" We looked up to see a white-haired gentleman dressed in a black suit addressing us.

"We were just admiring your little cat," Andrea offered as an excuse for our lingering at his front gate.

"Are you French, Swedish, British?" he asked in English.

"Well, we're American, and we were just out for a walk

and were enticed by your little cat," I said.

"Please come in for a visit. I am absolutely alone, absolutely alone. Do you have time to come in for tea?" he blurted out the gracious invitation.

So began our long friendship with Senhor João De Mello, former military officer, relative of a prime minister, member of the noble De Mello family, father of three, and a widower now for four years. He became our tutor in Portuguese life, our talking historian, our teacher of culture, but most of all, our dear, dear friend. This first visit was one of many that followed.

Upon entering his house, we realized the aloneness to which he referred was because of his widowed status, for his home certainly was not empty. At his attendance were a cook, a gardener, and a secretary/chauffeur. But socially he was by himself, and he was therefore delighted to make our acquaintance. We became his audience as he showed us his garden, told us stories of his childhood when he had romped with the royal princes and princesses of Portugal, shared his old photo albums, and regaled us with the best Portuguese food. His cook's chocolate mousse was legendary.

No matter that he repeated himself over and over. It was a pleasure just to hear him retell the stories that gave us indelible pictures of Portuguese life that would never be erased from our memory. They helped us to understand the Portuguese and their lifestyle better than we ever could have otherwise.

We also soon learned that "come to tea" in Senhor João's mind was more like the British "high tea" that we had heard about. Biscuits with spreads, bread and sweets, and other items appear on the table at teatime. To take tea in this fashion could also be expanded to mean a full-course meal, as it often did at Senhor João's.

"You must come visit me in my summer home at Sintra," he told us one day in late spring. "It becomes so stifling here in the city that we always go up to the mountain to have the cooler air." He explained that this was a traditional annual migration dating back to royal times when the king and leading families would seek higher ground to avoid the sweltering Lisbon summers.

Sintra Mountain, less than an hour's drive from downtown, was indeed Lisbon's paradise retreat. Several old royal castles adorned the mountain. Large manor homes, public promenades, and lush gardens surrounded the town of Sintra. As an added divine touch, at the foot of the mountain on the side opposite Lisbon, white sandy beaches at the Atlantic Ocean shore glistened like diamonds. While the oldsters relaxed in the cool shade at the top, the grandchildren and great-grandchildren headed for the beach.

We rang the bell at the estate entrance and waited for a maid in a white dress to open the creaking ivy-covered gate. She pointed to a house half hidden from our view by flowering shrubs and oak trees. We drove down the winding gravel drive and stopped before an impressive ancient gray house. Our eyes fell upon the central point of the scene, an ornate gurgling water fountain in the middle of a large gravel parking area.

"Welcome, welcome, I was afraid you would not find me," Senhor João greeted us as we climbed out of our car. As usual, we first had to be given a tour of the area. He walked us back up the drive to the house of his sister, the countess of Cartaxo. A slim female version of Senhor João, she was most gracious and witty. They talked of the old days when they were young and their families spent the summer months here in Sintra. At the death of their parents, she had inherited the main house of the estate, and João got the stables, which he promptly turned into a magnificent family home of his own.

"Come, I've arranged dinner for just the three of us," Senhor João urged. "All the grandchildren have gone to the beach, and my son and daughter are still at work in the city, so we are very much alone."

When we entered the dining room, we gasped as we looked at a linen-covered table long enough to serve dozens of people at a banquet! We were seated at one corner and began another pleasant conversation about the weather, the old days at Sintra, and the latest happenings with his family. We had come to know most of his immediate family, so we were able to make sincere inquiries about his son the banker or his son-in-law the professor or their sons and daughters, who were mostly students at that time.

To say that our friendship with Senhor João was a treat would be an understatement. For us, he represented Portugal in its fullness. For him, we were his surprise friends during his old age. He depended on us to make life meaningful at a time when he needed friends the most and his world was changing at a confusing rate.

After we left Portugal, we continued our friendship by letter, until one day an envelope arrived, written by an unfamiliar hand. It was from his daughter, informing us of his passing in 1979 at nearly ninety years of age. She thanked us for our friendship, which she said had meant so much to her father in recent years. For our part, Andrea and I could both say, "Senhor João, the pleasure was all ours."

Forever after in our minds, Senhor João epitomized the loving people of Portugal. We found that this same love and concern were shown on all levels of society. The graciousness we found in him, we found in the hundreds of dear church members who became friends for life. The hospitality we found in his home, we found in all Portuguese homes to which we were invited. The steadfast friendship he showered upon us was the very same shown to us by numerous people we came to know and love.

These lasting traits have reached out to us over the miles

and over the years, inspiring me to declare the Portuguese among the most loving and welcoming people in Europe. It's probably that same germ of love, transferred to another place and culture, that makes me think that the Brazilians are the most loving people of the Americas.

But when we first arrived in Lisbon, we were strangers in a foreign land. We did not yet speak Portuguese. In our early isolation, we spent many Saturday nights snuggled around the little propane space heater in our apartment, reading books.

One of the first important activities for any Adventist moving to a new location is to find a church home. We started visiting the Adventist churches in Lisbon one by one. Our hearts were captured by one in the suburbs that was full of young people and led by a dynamic pastoral duo. Pastor Pires and his wife, Maria Augusta, welcomed us with open arms.

"Does it snow in America?" "Do you know any American Indians?" "How big is New York?" were some of the questions the young people of the Amadora church bombarded us with in their halting English. We couldn't help but love these vivacious young people. Soon we were taking them home for Bible study and cookies on Friday nights, hiking with them on Saturday afternoons, and socializing with them on Saturday nights. Soon I was asked to be youth counselor for the church, and both of us joined the choir and became totally immersed in church life.

Two years after we arrived, Pastor Pires was hospitalized for an illness whose symptoms were compounded by injuries he had suffered in an automobile accident some years before. His situation became serious, and within weeks he died. Our joy in the little church family became our sorrow, shared with the Pires family and the extended family of the church.

One of the most profound moments in my life came as we, a corporate church family, were proceeding to the cemetery where we would lay our good pastor until the resurrection morning. Joining in their grief, Andrea and I plodded down the road following the coffin held aloft by young men of the church. I felt an arm take mine. One of the young men was pulling me forward so that I could take the pallbearer position. With heavy heart I bore the burden of the coffin with five others. From then on, I had no doubt but that I was a valued member of the church family. How could one become more integrated into a culture than to be invited to share in so intimate a service?

"We feel the Lord has a special work for us to do in the city." I tried to explain our decision to change churches, after three years at Amadora. The members were having difficulty accepting that we were actually going to leave their membership.

"Brother Steele is right." I heard the approving voice of Maria Augusta, who was now, after the death of her husband, pastoring the church. "He's right. We here at Amadora have active young people, a full church with standing room only, a great church family. The central church has young people but no leaders. Their pews are full of old people, and they need our help. Even though we will miss our beloved sister and brother, we must do this to help the central church revive its youth activities." Her words were accepted as final.

Our first priority at the central church was to start a junior Sabbath School. We were able to convert a downstairs classroom for our use, and soon it was packed every week. The children liked the idea of having their own place and program.

Next, a bigger task by far, was the formation of a Pathfinder Club, the first in Portugal. We invited children in the neighborhood and started the club right off with fifty young people, half of them not from Adventist families. Within a year, several were baptized. One, Maria do Rosario Raposo, became like a daughter to us five years later when she came to live with us while she studied at Andrews University.

One boy became my own personal burden because he was always talking out of turn and goading the other boys into acting the fool. Often when on a camping trip, he would bring me to my wit's end. I would send him to sit up in a tree, there to chirp like a bird, which I said he was because he couldn't keep his mouth closed for two minutes! Twelve years later, when we returned to Lisbon for a visit, he insisted on talking to me. He wanted me to know that his Pathfinder experience changed his life—he is now teaching physical education to high schoolers! Such are the rewards when working with young people.

"Brother Steele, would you lead out in our first junior youth camp?" the church youth director asked a few months before we left Portugal. At the time, I was desperately trying to wrap up the radio work so I could reach Andrews University in time for classes.

"Well, I can't possibly lead out in the camp, but I am willing to attend and help out," I replied cautiously.

It was a glorious late summer, and we drove for four hours to reach Figueira da Foz, where the church camp was located. The open-air gym/cafeteria sat back from the Atlantic beach several hundred feet. Tents were to be our home for the week. I arrived just before the scheduled staff meeting was to begin.

"We are assembled here for the first junior youth camp in the history of the church in Portugal," said the youth leader. "And I would like to present to you the director of our camp, Allen Steele." I was in shock! I had never agreed to be camp leader, but it would be embarrassing to reject the job after such an introduction. Fortunately, we had a very good group of camp counselors. The week would be possible, I determined.

The days were filled with camp crafts, swims at the beach, and games. At night we showed films borrowed from several embassies in Lisbon. As the weekend neared, my thoughts turned to Fernando. He was a conscientious young man in his early twenties who had brought his little sister to Pathfinder Club but became interested in the activities himself. Soon he was a regular counselor. I had studied the Bible with him, and I felt his heart was open to God's leading.

"Fernando." I spoke to him one evening as we stood at the edge of the gym watching the Pathfinders play a game of steal the bacon. "You have been with us now for over a year. You obviously agree with everything we teach, and you've been faithful at all our meetings. How would you like to be baptized here at this camp, this weekend, as a witness to these young people?"

"I think I would like that," he said in his nonchalant way. His baptism was the climax of the camp, and many Pathfinders indicated they wished to follow his example.

Chapter 7

"Can You Stay Another Year?"

"I can see it now," I yelled to Andrea above the traffic noise. "It's a Pan Am 747. See that little white dot?"

We were at Lisbon's Portela Airport awaiting the arrival of our first visitor from the General Conference, Walter Scragg. Shortly after we moved to Lisbon, Walter was appointed director of the radio-TV department when James Aitken became the church's representative to the United Nations. At the same time, Harold Reiner became associate director of the department in Walter's place.

Our most recent correspondence from Walter was not encouraging. He said that the AWR financial situation was serious and that we might not be able to continue our work in Portugal. We were very anxious about the news he would bring.

"I have good news for you," he said as we piled into our VW minibus for the trip to the office. "Things have turned around. After I told you we were low on funds for AWR, we worked on a little report for the back page of the *Review*. We mentioned the challenge we have to keep the radio programs going, and within two months, enough funds came in to cover next year!" He gave us his big Australian smile and added, "Isn't God good?"

We sighed in relief. We couldn't bear the thought of cutting our work short. Our program producers were doing well at keeping the programs coming to us. New studios, in Sweden, Holland, and England, were about to begin production, and we had received one thousand letters during our first six months on the air.

We could tell that Walter shared the relief too. What he told us about the result of the article in the *Review* became a pattern for the first few years of AWR's life. When we faced a shortfall of funds, Walter would crank up his writing arm and get an article in the church paper, and miraculously, readers would respond, and funds would flow. Each report gave AWR another grasp on life.

"I've been authorized to ask if you two are willing to stay another year in Portugal," Walter said when we arrived at the office.

"Of course, we are," we both chimed. "We feel we've only just begun."

It was great fun to have a visitor. Not many church leaders came to Lisbon, since it was not on the way to many of Europe's major cities. As a matter of fact, we seldom, if ever, saw Americans in Portugal, although British tourists were quite common. Each visitor to AWR was a boost to our morale and gave us an opportunity to learn more about the country as we toured with them.

A few months after Walter's first visit, another important couple flew to Lisbon. A Swedish-American, Dr. Olov Blomquist, and his wife, Willma, were excited about the potential of our station in Portugal to reach their homeland in Scandinavia. They were interested in every aspect of our work, and we soon realized they were real kindred spirits.

Olov and Willma became father and mother to us fledgling broadcasters and many others. We always knew they had a sympathetic ear when we were discouraged and loads of good counsel when we faced challenges. Many broadcasters, in time of discouragement, found solace and new life from the enthusiasm and concern of the Blomquists. They found funds for many new radio studios and stations and were always keen to nudge the church into the broadcast ministry whenever and wherever they could.

The on-the-air studio in Sesimbra and the transmitter site in Sines were required visits for anyone coming to see AWR at work. Both these facilities were operated and maintained by Deutsche Welle, the German international broadcast organization. Radio Trans Europe was privately owned, but with Deutsche Welle as a partner, technical and engineering aspects of the operation were expertly cared for.

The trip to Sesimbra was a picturesque drive. The first segment was over the Salazar Bridge, which crossed the Tagus River. When built, the bridge was named after the country's ruling dictator, then later renamed the "April 25 Bridge" when democracy was declared in 1974.

From the southern side of the Tagus, there was a stretch of about fifteen kilometers of Portugal's only express highway, save for a similar short stretch north of the city. Upon leaving the expressway, the remaining section of the journey was across the high cape of Cabo de Espichel, dotted with small farms and herds of sheep.

This last section of the trip was the most pleasant. I was glad that my weekly trip to Sesimbra was blessed with such pleasant scenery. Once, I stopped near a little stone barn just to see what it contained and how it was built. As I rounded a corner of the whitewashed building, I gazed upon a most tranquil scene.

A big white sow lay on her side, snoring with a dozen little pink piglets stretched out between her legs. I quietly took my camera from my shoulder to record this peaceful scene. The shutter sound awakened a little black-and-white kitten I had not noticed. Upon waking, the little tiger began hissing and spitting with great vehemence.

This awakened some of the piglets, and pandemonium broke loose. Pigs and cats were jumping in all directions. I made a hasty retreat, happy that I had the scene recorded on film.

A narrow, winding road descends precipitously from the cape to the white sandy beach where the little fishing village of Sesimbra is situated. In summer it is a holiday resort, but for most of the rest of the year it is a somnolent village—except when the fishing boats come in. On these occasions the whole beach becomes a giant fish market.

It is near this picturesque town that the Radio Trans Europe studio was built so that it could have a direct line of sight for microwave links with the transmitter site in Sines, some fifty kilometers farther south on the same coast.

The trip south to Sines is through one of Portugal's main cork forests. It is a special experience to see these trees, members of the oak family. Their trunks, stripped of the cork bark, reveal a range of gold, orange, and brown color, depending on how long it has been since they were stripped. Generally, cork is harvested once every seven years. Portugal is the world's leading producer of cork.

The town of Sines, during the days we lived in Portugal, was another sleepy fishing village. On days when there was no ocean breeze, the air was stiflingly hot. At midday the whole town came to a standstill for siesta. The small man-made harbor accommodated a dozen colorful fishing boats.

This was the birthplace of Vasco da Gama, the great Portuguese explorer. A few kilometers outside town is a small hill named Monte Mudo (Silent Mountain), which became home to the shortwave antennas of Radio Trans Europe. Four 250-kilowatt transmitters, three used by Deutsche Welle and one leased by Radio Trans Europe to organizations such as AWR, were housed in a modern, air-conditioned building surrounded by a small, well-cared-for garden of flowering plants.

The silence of the place always overwhelmed me. Two sounds come to mind as I remember the times I stood out near the antenna field. The gentlest winds caused a rushing sound when they passed through the antennas, stretched up against the sky. When I heard that sound, I prayed that the winds would carry God's message swiftly across the mountains and valleys to homes where eager ears waited to listen.

The other sound that usually floated through the air was the sound of a braying donkey from a nearby farm. The little domesticated donkey, or burro, is the Portuguese farmer's mainstay for work and transportation.

These two facilities at Sesimbra and Sines combined to constitute a powerful tool for declaring God's message of love and salvation to more than five hundred million people in Europe plus North Africa and the Middle East.

By the time we reached our second year with AWR, many stories about listeners had reached our desk. From every corner of Europe, people wrote to tell of their experience in discovering AWR programs. One day, when the Blomquists were visiting, we went through the recent mail and reviewed some of the most exciting stories.

"Sweden must be a big country," Andrea said as she told about a young woman living near the Norwegian border. "She says she had never heard of Seventh-day Adventists before she heard our broadcast, and when she finished our Bible correspondence course, she didn't know how to find an Adventist church. Finally, a pastor from a church two hundred kilometers away was able to come by train to visit her. He baptized her, and now once a month she travels by train to church."

"And what about this man in Germany?" I suggested. "He listened to our program, signed up for the Bible course, and decided to become a Christian. But which church? At the same time he was studying our Bible course, another Christian was visiting him weekly to study the Bible. He felt it was time to make a decision, so he prayed and asked God to send an Adventist to his door the next day if that was the path he was to follow. Next day, an Adventist pastor who had been alerted to the man's needs by AWR went to visit him. He accepted it as the sign he had asked of God. But there's more to the story," I added. "He is partially blind and is now helping to translate Christian literature into Braille for our Bible School in Darmstadt."

"Wait till you hear this one," Andrea said. "A Swiss family in the mountains started listening to our programs and decided they, too, wanted to become Christians. Only one problem. The family owned a tavern, and they knew they couldn't in good conscience continue this business." The high point of the story now wavered somewhere on the highest cliff of the Alps.

"The little six-year-old-girl in the family came up with a terrific idea," Andrea explained. " 'Let's turn our tavern into a milk bar,' she suggested. 'We can sell milk, cheese, ice cream, and yogurt.' Well," Andrea continued, "that's just what the family did, and they report that business is booming."

"Here's a family of six who were recently baptized because they listened to AWR." I thought this would please Olov and Willma because I had a picture of the family on their baptismal day. "They are Ukrainians who were able to emigrate to France. Being far from home, they listened to the radio to see if they could find a Ukrainian station. What they found was the Ukrainian *Voice of Hope* on AWR. They identified with the message and enrolled in the Bible course."

"Well, here's a story you'll find hard to believe." Andrea just had to have the best story of all! "A member of the Communist Party in Russia started listening to AWR and decided he wanted to take the free Bible course that was offered." Her eyes began to sparkle as she prepared to divulge the secret details of the story.

"He knew that he couldn't openly receive the Bible lessons through the mail at his office or home address. So he began to think of how he could get the lessons without anyone knowing. Three times a year he had to travel to the Russian Embassy in Rome, and so he thought, *Maybe I can have them sent somewhere in Rome.* He decided to ask his barber in Rome if he would receive the lessons for him and save them for his regular visit. And that's how this man is now studying the Bible course!"

Olov and Willma now had tears streaming down their cheeks as they contemplated the power and miracle of radio. Stories like these multiplied over and over through the years. During those precious moments when we were with the Blomquists at the annual AWR board of directors meeting, we stopped for "AWR storytime." The stories fueled their fire to give the world the gospel by radio, and it always gave us AWR staffers a chance to be thankful for what the Lord was doing through the radio ministry of the church.

Chapter 8

Secret Christians

"It's hard to believe we're actually going to visit eastern Europe," I said as Andrea and I discussed our upcoming trip to Yugoslavia. We had heard from Roger Fasnacht and Bernard Pichot about the fascinating way they were able to bring people together once a year to record programs in Serbian, Croatian, Macedonian, Slovenian, and other eastern European languages, and now we would witness this spectacular event firsthand.

"Are you sure we can get through the border OK?" Andrea seemed to want reassurance.

"According to tourist information, you can get your visa at the border," I answered. "But sometimes you don't know the truth till you get there."

We worked hard at the office to get our work prepared for the days we would be gone, then set out in our little blue Fiat 124 across Spain, through the Riviera of France, and on across northern Italy to the Yugoslav border. Our source had been correct, and we were able to buy a visa at the border and travel on through a most beautiful countryside. We were entranced by the pastoral setting and the active farming, with farmers utilizing large horses, nothing like the little donkeys of Portugal.

It was late when we neared the Adventist college at

Castle Marusevec, near Zagreb. The castle came into the church's hands when Prime Minister Tito's government ordered Adventists to move their seminary out of Belgrade. In return, the government allowed the church to establish its school in an old baronial castle in northern Yugoslavia, provided they would commit to restoring and maintaining it as a historical site.

We were trying to follow the map in the dim evening light. We found the turns that were indicated but were surprised to be headed down a dirt road. We stopped to pore over the map and immediately found ourselves in the dust of a horse-drawn hay wagon that had come upon us from behind.

"Marusevec, Marusevec," we shouted to two stout farm women walking beside the wagon. They replied with words in their own language and pointed in the direction we were headed.

"Welcome, welcome, please come in for supper," said the school principal when we finally arrived.

A yogurt soup with solid brown bread tasted good. Hot tea and biscuits filled in the corners of our stomachs, which had not had solid food all day.

The most exciting part of our visit came the next day. We found that here, in the southern woods of Europe, nearly one hundred people had come to take part in the annual radio-recording session. From sunup till late into the evening, our recording technicians kept at the work of recording speakers and musical groups in five different languages. At the end of the two-week session, they would leave with hundreds of tapes filled with material for the next batch of programs for AWR. What a labor of love!

On another trip, we were able to travel into Czechoslovakia and Hungary to experience life in the East. We sympathized with the pathos of the people burdened by Communist rule.

We listened as our guide in Prague derided the Russian

soldiers who had been sent to "protect"the Czech citizens and were posted around the city. We pitied the family in Budapest who moved out to sleep in tents in the backyard so they could rent their house to us for desperately needed income. Then we were distressed when the housewife asked us for used clothing, and we had nothing to give her because we were traveling with only enough clothing to make it through the trip.

We were intrigued by the black-market money-changers in Bratislava. The hotel where we stayed had two cashier drawers: one for government transactions, the other for black-market money exchanges. We laughed at the ice-cream cones we bought in Hungary. They felt like frozen creamed sand in our mouths.

When we finally returned to Austria, we stopped for a meal of steaming noodle soup, wonderful fresh bread, omelets, and salad. It seemed like a banquet after the meager offerings across the border in the East. We laughed because our shrunken stomachs didn't seem able to hold all the food of that meal.

"Do you think Communism will last forever?" Andrea asked as we traveled through the countryside of Switzerland. "Do you suppose our broadcasts through the Iron Curtain will ever bring about a change?" She asked the questions as if thinking out loud, rather than expecting an answer. Our experience in the East made us all the more determined to continue our work of penetrating the Iron Curtain with radio waves.

Another curtain that we knew existed was the curtain between Europe and Africa. From time to time, word would come out of those lands about people responding to the Adventist message through our Arabic broadcasts produced for North Africa at AWR's Paris studio.

In the first four years, we received letters from twenty-seven countries in response to the Arabic broadcasts, and the Bible School had five hundred active students. Pastor

Pellicer, founder of these broadcasts, kept very busy running the Bible School, preparing programs, and personally visiting interested people in North Africa. He shared some very heartwarming stories with us.

"A young man wrote to the Bible School in Paris to say he was a long-time listener to AWR, and he wished to take the Bible course," he told us. "I could tell that he had an unusual sincerity about this subject, so I decided to send him the whole set of Life of Christ lessons. Within two weeks, all the completed lessons came back to us in the mail. Then we sent him the book *Steps to Christ,* a Bible, and some other literature. A constant stream of questions indicated to us that he was delving deep into the Adventist message. The school was able to send him the address of an Adventist church. Shortly thereafter, the Bible School received the names and addresses of eight more young men whom this fellow had convinced to take the course. Needless to say, this young man became a church member, as did several of his friends.

"A young woman in Egypt fell into conversations with her Coptic neighbors about Bible topics," he related as he started another story. "She had been given a Bible-course enrollment card on the streets one day, and when she applied for the course, she was informed of the radio programs. Because of her new beliefs, she was thrown out of the house by her parents. She remained true to her newfound faith, however, and was soon baptized.

"Another young Muslim, a teenager who was known as the brightest lad in his class, heard our broadcast and requested the Bible lessons. He became so excited when the first lesson came that he decided to take it to school with him, hoping he would have time to work on the lesson. He was so anxious to start filling out the lesson that during one class, every time the teacher turned to write on the blackboard, this lad opened his briefcase and tried to answer the Bible questions. But once he was not quick enough, and the professor saw him looking in his briefcase.

"The professor walked back to the student's desk, opened his briefcase, grabbed the Bible lesson, and said, 'You stay after class.' The boy was frightened, but after class, the professor looked over the lesson with him. The teacher became interested in the lesson, too, and asked if the student would mind sharing the lesson study." Pastor Pellicer slapped his knee and said, "And that's how they both became our Bible-course students."

Obviously, Pastor Pellicer was dedicated to his work and confident that the radio programs would bring many people in Arabic lands to Jesus. "Of course," he warned, "people must be very careful. To become a Christian is very often the same as becoming a dead man," he emphasized. "It is contrary to the Muslim culture, so most Christians down there must be secret believers."

One of the spiritual highlights of 1975 at our seminary in Collonges, France, was the emotional baptism of a young Arabic man whose first contact with Adventism was through the Arabic Bible course. In giving his personal testimony, this handsome youth told of his search early in life for something fulfilling.

"My parents were able to steer me clear of many obstacles such as drinking and smoking," he said in his testimony. "But when I was twenty, some friends of mine tricked me into taking some drugs with a soft drink. This led me into the drug scene." He became very quiet and took a deep breath.

"Some months later, I realized what a terrible path I was following. I was regretting these steps and decided one day that I would go back to my old hobby of stamp collecting to occupy my time.

"A friend told me that the Voice of Prophecy Bible School sent stamps to anyone requesting them, so I wrote to the Bible School and asked for stamps and at the same time signed up for the course they offered. This led me into a new spiritual experience, and I decided I would go to the Adventist school in Collonges." It was there that he was baptized and welcomed into the Adventist Church.

A story similar to this was repeated some years later. A young man listened to our programs over a period of several months and became convinced that the Lord was calling him to Christianity. He could find no Adventist church in his city in North Africa, but he did find a small evangelical group, which he joined clandestinely.

He wanted to be baptized as an Adventist but accepted baptism by this group, saying in his heart that he was being baptized a Seventh-day Adventist. About that same time, he fell in love with a lovely young woman and immediately faced the dilemma of figuring out how he could tell her that he had become a Christian.

He began by telling her about Christian beliefs that a woman would like to hear: Christian men marry only one wife, marriage for Christians is a lifelong commitment, and women are considered equal to men in the Christian home. Before long, he had her convinced that Christianity was the right religion—*and* that she ought to marry him! Not long after that, he went to study Adventist theology at Collonges, where he completed his course as a religion major.

Church leaders had their eye on him and invited him to become a radio speaker for *The Voice of Hope*. Andrea and I asked him, once he accepted the job, how he planned to attract his people to the Christian message.

"I will tell them about God's love," he said, "because it is something they don't know about or understand." He went on to become a radio pastor to thousands of secret Christians in Arabic lands.

Such stories make us long for the day when we shall meet face to face with those thousands of people who must follow their convictions secretly and suffer the loneliness of being outcasts in their own families and society.

Chapter 9

Trials and Triumphs

"You won't believe what I had to go through to get this bottle of cooking oil," Andrea said with a hint of distress. She had obviously gone through an ordeal, for she seemed tired and irritated.

"I asked for oil at several stores and was finally directed to a little shop down Blind Alley," she continued.

Blind Alley was the name we had given to a very narrow street a few blocks from our apartment in Alfama. It was only wide enough for one car or one streetcar to pass through at a time. When riding the streetcar, you could actually reach out the window and touch the building fronts!

It was a place that every driver hated but had to use if he or she wanted to get to our section of town. City hall tried to help the situation by stationing two watchmen as directors of traffic. Each had a "Ping-Pong paddle" with red paint on one side, green on the other. Their job was to sit at the only spot where traffic from both directions could be seen and give the red or green signal, depending on which way was clear.

The plan worked quite well unless one of the watchmen left his post for a few minutes or fell asleep on the job or got embroiled in an argument. Then vehicles from both directions would enter Blind Alley, and at the middle both would claim right of way. A snarling traffic jam would ensue, horns would blare, tempers would flare, and it would take hours to clear the alley. We learned to accept it as part of the color of Alfama.

"Then I had to wait until there was no one else in the shop before I could ask for the oil." Andrea's voice brought me back to her story. "Then I had to assure the shopkeeper that I would become a regular patron of his store. Then I had to pay twice as much as normal for the stupid oil. Then I had to wrap it in paper so no one could see it," she said with finality, close to tears. I knew there must be more "thens" to her story, but she was too frustrated to go on.

"Well, you know it was the same way at the service station," I tried to console her. "The line of cars wound around the block. I had to wait a full hour to get to the pump; then the station attendants announced they had just run out of gasoline, and I had to start the process all over at another station!"

That was the way life was in Portugal in late 1973 and on into 1974. People were short-tempered and irritable. Food shortages were common. We heard of hostilities in the overseas provinces of Angola, Mozambique, and Timor. We wondered what it would all lead to.

It led to the Portuguese Revolution in April 1974. The events of the Flower Revolution lasted only a couple of days, but the results of the change of government affected Portugal for many years. The euphoria of the victory soon turned into disillusionment fueled by broken promises, failures of political parties to deliver, and economic turmoil.

With the revolution complete at home, the military leaders turned their attention to the overseas territories. Their solution to those problems was to grant the countries immediate independence. Portugal could no longer afford to maintain its sovereignty over them, the military declared, and besides, the territories no longer appreciated Portugal's rule. Thus the world's last great colonial power relinquished its control over millions of people in Africa and Asia.

Suddenly, Lisbon and the rest of Portugal were overrun by *retornados*, Portuguese citizens who were forced to leave their overseas jobs and homes, many with just the clothes on their backs. Bankers and business people became paupers overnight. Beggars began to appear in Lisbon's subway stations and on street corners, something rarely seen before the revolution. The city was filled with people in shock. Hotels were forced to take in the refugees for temporary shelter, the rooms paid for by foreign governments that were asked to help.

"Where's our car?" I asked as Andrea and I left our apartment one Saturday morning, all dressed for church. "I know I parked it on the other side of the street!" I said in dismay, looking at the empty spot where I left the car the night before. We quickly looked up and down the street, but no car.

We walked ten blocks to the nearest police precinct to report the missing car, then halfheartedly walked home to contemplate our latest dilemma. The police officer said we would be informed when the car was found, but how effectively could they look for a stolen automobile? Some of our church friends, when they heard of the theft, said the best way to find a stolen car these days was to ask the radio stations to announce it on the air; then listeners would be on the lookout and call the station.

Before the weekend was over, there had been four sightings of cars fitting the description of our red Fiat. But having no transportation, we had no way to follow up the reports.

A week later, a police station called to report that our car had been spotted by a guard at the police post. The alert guard watched as the car passed his post, then was parked in a nearby square. With the help of a colleague, he went to inspect the car and found a man sitting in the front passenger seat. The man was immediately arrested.

The car looked as though it had been driven through the woods at night; scratches on both sides revealed the mistreatment. It was also out of gasoline, and the key to the gas tank was nowhere to be found. I used a hacksaw to cut off the locked tank cover. Then I hiked to a service station several kilometers away to get a gas can filled.

When I returned with the full gasoline can, the car was gone! In disgust, I went back to report the missing car again to police and headed home to wait for any news.

"It's the police station; they've found the car," Andrea said as she handed me the telephone when I walked in the door. "They say they found it abandoned in a square and had it towed to the main station."

I immediately caught a taxi and went to the main police station to collect the car. We were glad to have wheels again, and with a new paint job, it looked like new. Several months later, I was asked to witness in court that it was indeed my car that had been stolen. The man the police had found in the car was acquitted because the judge said the police could not prove that he was the man who stole the car; he was only a passenger.

Through these trying times, we were again able to praise the Lord that our work was not disrupted. AWR continued its broadcasts with good success. The mail count increased year by year, and by the end of the third year, we had received approximately ten thousand pieces. New stories of conversions continued to come to our office.

Erwin Killian, communication director of the Euro-Africa Division, reported from Germany that a forest ranger was going to his post early one morning with his shortwave radio over his shoulder. Suddenly, his attention concentrated on what was being said on the radio. For the first time, he was hearing an AWR program, and the message gripped his heart. That led to his sending for the Bible School lessons, then to personal study with the local pastor, and finally to his baptism on the last Sabbath of 1974.

One of the AWR speakers in Yugoslavia received a letter from a young man who had heard our program, taken the Bible course, and now was requesting baptism. Consulting his busy schedule, our speaker wrote the listener, telling him that, unfortunately, he would not be able to come to baptize him for several weeks.

When our radio speaker did finally arrive in the city, he found to his surprise that twelve people were ready for baptism! The young man had been busy with those Bible lessons!

By the end of 1974, we had added Turkish and Norwegian programs to our schedule. This brought the total number of languages to seventeen. In 1975, the number of languages planned climbed to twenty. Another funding crisis came at the end of 1974, but Walter Scragg was given space in the *Review*, and donations increased. Still, the outlook for 1975 was grim.

In 1975 Andrea and I had the privilege of attending the General Conference session in Vienna. To help get reports from the conference on the air and to have someone at the office while we were away, we got permission to hire a helper. We found a capable person in Manuel Vieira, a young married man who was interested in learning the radio work. His talents and interests were such that he continued with AWR as a valuable employee for many years thereafter. In 1995 he became the first person to receive a twenty-year service award from AWR.

The quinquennial business session of the worldwide Adventist Church was especially thrilling because it was the first time ever that delegates from Bulgaria were permitted by their government to attend. The Russian church also obtained permission to send a delegation for the first time in forty years.

The elections at the meetings brought about leadership changes for AWR. Walter Scragg was elected president of the Northern Europe–West Africa Division, headquartered near London. M. Carol Hetzell, a long-time journalist in the church, was elected director of the new communication department at world headquarters, which incorporated the radio ministry.

We departed Vienna at a point when politics were boiling in Portugal. Clashes between the Communist Party and the rightist parties were at a fever pitch. The Communists were able to bring in land reform in the agrarian south of the country, but the small landholders of the north, who felt they had already been independent for decades, resisted these reforms. More than resisting, they burned down the Communist headquarters in the north.

Press coverage of the isolated clashes in Portugal gave the world the impression that the whole country was ablaze. Relatives and church workers were concerned for our safety.

At the Vienna meetings, the church treasurer from division headquarters in Bern, Switzerland, was concerned about our return to Portugal. Stanley Folkenberg and his wife, Barbara, had become like parents to us and took personal interest in all that we were doing at AWR. We assured them that we felt perfectly safe in Lisbon.

In late 1975, a new opportunity for AWR presented itself. The same businessman who had arranged air time for us on Radio Trans Europe said he could also get air time for us on a new radio station on Malta, with good Middle East coverage, and also on a station in Kigali, Rwanda, to reach into Africa. I visited Malta to investigate the possibilities there.

"I didn't want to arrive in Malta this way, by airplane," I wrote for the *Review* in the summer of 1975. "I would much rather have traveled by boat, like the prisoner Paul,"

I explained. To be on the island at all was a thrill, and I was especially moved when I visited St. Paul's Bay, where the apostle is supposed to have landed.

The new station at Delimara Point had an impressive antenna array. On August 1, 1975, AWR broadcasts began on this station over a 250-kilowatt shortwave transmitter. We had hopes that we eventually would also be permitted to utilize a 600-kilowatt medium-wave transmitter, but that never happened.

Plans for new programs in Africa were scuttled when AWR funds dropped precariously at the end of 1975. In fact, the situation became so serious that church officers feared AWR would face a complete close down. Fortunately, church leaders at the General Conference and at the European division offices came to the rescue with additional funding. However, a tighter budget required that three languages—Spanish, Norwegian, and Macedonian—be dropped from the schedule, and broadcasts were reduced to eleven hours a week.

"If I don't go back to school next year, I'll miss out on my veteran's educational benefits from the government," I told Andrea one day in late 1975. "There is a deadline for applying for the financial aid, and mine is next year."

"Well, I know you enjoy your work here at AWR," she said, "but the longer you wait, the more difficult it will be to go back to school. Why don't we ask for a one-year study leave?" she suggested.

I appreciated her support and understanding, and her suggestion was a good one. We decided to try it.

"Here's a letter from the GC secretariat." She came into the studio where I was working several weeks later, holding a white envelope in her hand. "Hurry and open it; it may be our permission for a study leave."

She was right. We were being granted a study leave with the hopes that we would return to Europe in a year's time. On the recommendation of Olov and Willma Blomquist,

Ron Myers, a popular gospel music announcer in southern California, was asked to come to Europe for a year.

Our five years in Portugal had been exciting ones for us and for AWR. The radio ministry now seemed to be on a strong footing. The next few years would turn out to be incredible years of expansion for this infant work.

Chapter 10

AWR Diary

"Aren't you a pretty sight!" were my first words of greeting to my wife at Andrews University. She had traveled to Berrien Springs, Michigan, a few weeks before I was able to leave my responsibilities in Lisbon. Now I found her in the middle of our friend's kitchen, hands and arms covered with applesauce.

"Yes, I'm a mess, but won't you be glad for this applesauce next winter?" was her quick reply. We found that canning fruits and vegetables for the long cold winter was the universal summer activity for people in this part of the United States.

We had a joyous reunion, quickly found an apartment, and were settled in, ready for the academic course and winter ahead.

"Welcome to Andrews University," said Dr. Sakae Kubo, my advisor for my master's degree studies. Then he surprised me with a question, "What classes to you want to take?" I had expected him to tell me what courses I had to take, so to be invited to choose my own plan of study was incredible.

"Well, in the religious communication course, you simply choose half your classes from the theological seminary offerings and the other half from the communication department," he explained.

Soon I was indeed a student again. The winter weather encouraged academic pursuits, and I liked the quarter system because it gave me more opportunities to begin anew in various fields of study. My year of study for the master's degree was full and went fast.

I was so busy studying that I had little time to think of radio. I did visit the university radio station, WAUS, a few times, and one thing that intrigued me was the fact that the station was utilizing satellite feeds from National Public Radio for much of its programming. I began to think about how wonderful it would be if the church could someday take advantage of the same technology to transmit its programs to stations throughout America and around the world.

As my graduation with a master's degree neared in the summer of 1977, I was called to the university president's office. "Would you be willing to stay on at Andrews to be manager of WAUS, if the university would help finance your study on the doctoral level?" Dr. Grady Smoot asked.

I had heard that an opening at the station had developed, but since I was committed to AWR, I never thought of myself as a candidate for manager. The offer of help with a doctorate was a carrot dangling before me. I immediately realized that the advanced degree might be valuable to my work at AWR, if they could do without me for a few more years.

"Ron is doing a good job and is willing to stay," Carol Hetzel wrote to tell me a few days later. "You are free to stay at Andrews so long as you plan to return to AWR someday," she wrote half seriously.

Thus began an incredible seven-year period of personal growth in study, teaching for the communication department, and managing the university radio station. During that time, I did everything I could to keep up to date on AWR. For my record of AWR activities during those years,

I depend on my news clippings and notes.

In the summer of 1976, Harold Reiner went to Macao to organize a studio to prepare Cantonese programs for a ten-kilowatt medium-wave station. Cantonese is the language of one hundred million people in southern China. For some years, Milton Lee had produced programs in Mandarin aimed at China from Taiwan. Now the power of radio would be tried on the southern part of this most populous country on earth! The task of making programs for the Macao station was taken up by Dr. Samuel Young, and he was soon able to report that "in the first three-and-a-half months the program received 600 letters, many from Mainland China, where people were hearing the church's message for the first time." We all hoped that someday we would have our own radio station to reach all of China.

In October 1976, the AWR board approved broadcasts under the name AWR-Asia over Radio Sri Lanka, to be prepared in the church's Poona (Pune), India, studio under the supervision of Adrian Peterson. Another religious broadcaster, Trans World Radio, was able to obtain full-time use of a big medium-wave transmitter at the same station and soon became the dominant Christian broadcaster to India.

In 1977 the Adventist radio pioneer H. M. S. Richards, Sr., came to the campus of Andrews University. In a visit to our radio station, he reported that the government of Liberia was willing to grant AWR a broadcast license. His wit and wisdom were absorbed by young seminary students who gathered to chat with him wherever he went on campus. For some reason, I remember one of his comments more than others. When asked by a seminary student what the greatest influence in his life had been, I remember the great man saying, "My wife, my wonderful wife." I felt privileged to interview him at that time because some months later, he passed away, leaving his wonderful Mabel matriarch of the Richards family.

That same year, the General Conference committee meeting at church headquarters accepted a long-range plan devised by the communication department to cover the world with the AWR signal. It would require three or four large shortwave stations. Funds were also approved to prepare a feasibility study for a station in Liberia, but the project never developed.

In Portugal, Ron Myers was caught up in the concept and continued studying possibilities to accomplish the goal set out for AWR. "I think the possibility of constructing an AWR station on the Portuguese island of Madeira is good," he wrote. "Manuel Vieira visited the island and spoke with the president, who was apparently excited about the idea," he went on. "The president has asked for a formal proposal from AWR."

Ron proposed a budget of $1,500,000 for a 250-kilowatt transmitter station with two curtain antennas. His calculations showed that the church could save nearly $100,000 a year if it owned and operated its own station, rather than leasing time, as we had been doing. His figures showed that the savings would permit thirty-eight additional hours of broadcasting per week on our own station.

Excited about the potential of reaching all the world, the Lisbon team of Ron and Annie Myers and Manuel Vieira prepared a document that proposed stations at Madeira, the Philippines (with Guam listed as an alternative), Seychelles, and the Bahamas. Ironically, Guam, the site they considered least likely, was eventually selected for construction of an AWR station.

Another letter quickly followed. "Did you know that FM radio is exploding all over Italy?" I could detect Ron's excitement between the lines. "Last year's test case against an Adventist businessman who was operating his own radio/TV station out of his house was won by the defendant. Now, hundreds of small-to-medium-size stations are popping on without licenses and are perfectly legal. I counted 20 or more just in Rome. For $25,000 or less we can plant an Adventist radio station in the heart of Rome." He ended with a plea, "Can we get even some used equipment together?"

With the help of the Blomquists and other friends, Ron was soon able to put an FM station on the air in Rome, located a few blocks from the Vatican. It seemed a miracle that such a thing could happen in Europe! The station has had an incredible history of ups and downs, including bombing of the studios, mysterious loss of equipment, and other tragedies, but today it is the flagship of a struggling but active network of local stations operated by the church in that country.

Italy was the first country in Europe to totally open up the airwaves to anyone who could broadcast responsibly, but soon other countries gave the same freedom: France, Belgium, Sweden, Norway, Denmark, and, in the 1990s, Romania. Our people in each of those countries could write a book about their pioneering experiences.

A story that revealed what happened in one country was told to me by Gosta Wiklander, communication director (and later president) of the Swedish Union. He said he was invited to a government-sponsored meeting. At the meeting, the national leader of the opposition party happened to be sitting next to him at the table. Gosta took the opportunity to tell him about the Adventist radio ministry.

"And why do you have to go all the way to Portugal to put your programs on the air?" the political leader asked ignorantly.

"Because the government will not permit us to go on the air here in Sweden," Gosta replied.

"That's ridiculous," said the man disgustedly. "When I'm elected, I'll see that that is changed."

Sure enough, in the next election, the opposition party came to power. Their leader was true to his word. Not only were religious groups permitted on local radio, but the government financed the building of most of these new community stations. Local churches were invited to share the air time with other community organizations. Usually, it meant two to four hours a day, in a block, for our church.

The new opportunities proved the true mettle of our church members. Usually Adventist programs only went on the air when individual church members were willing to make the personal sacrifices required to maintain the programming. This meant long volunteer hours, learning by mistakes how to produce programs, and struggling to locate resources. These volunteers are my heroes. In spite of little or no financial support from the organized conferences, unions, divisions, or General Conference, they make it happen.

Paolo Benini, in Italy, is one man who has done a lot for radio work. He was president of the Italian Union when I met him in 1993. Before coming to the top spot in leadership, however, he was instrumental in establishing several local radio stations in Sicily. As president, he was proud of the very effective radio station in Rome. Under his leadership, the station was able to increase its power, obtain a new transmitter and some new automation equipment, and become the church's best witness in the "Eternal City."

In the long haul, I observed, it is the heavy burden of creating programs on an hourly, daily, and weekly basis that causes local stations to fail. A satellite-distribution system can spread the load over more people and revive many struggling stations. In 1995, the Novo Tempo national radio network in Brazil became the church's first network to take advantage of this method of program sharing.

It was cause for great satisfaction to know that AWR had generated interest in radio as a way to reach people. Our shortwave broadcasts also demonstrated to the governments of Europe that Seventh-day Adventists had a message so important that we would develop any means necessary to make it available to everyone.

Along with good news from Europe came bad news, in 1977. While its big neighbor to the north was expanding individual freedom, the island government of Malta was objecting to AWR's plans to broadcast programs to Yugoslavia from the new Radio Mediterranean. The island government was apparently getting pressure from all sides. Now they were edgy about their relations with Communist eastern Europe. We had hoped this station would provide a way to broadcast to the vital eastern European and Arabic countries of the Middle East, but instead, it was only able to broadcast our western European languages. Finally, in 1982, AWR ceased broadcasting from this station.

With the decreasing possibilities on Malta, Ron began to investigate other possibilities in Europe. He found a small station in Andorra that was willing to take Adventist programs.

"With the booms of rocketing fireworks bursting outside the studio window, AWR-Europe's first-ever live broadcast began over Radio Andorra International," he wrote about September 8, 1979. "A hundred thousand or more visitors had come to help the thirty thousand native Andorrans celebrate their national holiday. But few of the festivity makers were aware of another important event taking place at that same moment," he said, "the first AWR transmissions on this small but mighty three-kilowatt station nestled in the Pyrenees Mountains."

Broadcasts from Andorra continued until the spring of 1981, when the government closed all stations in its territory. Concessions to operate had expired for private broadcasting, and the authorities nationalized all radio and television stations.

The AWR signal over Radio Trans Europe from Portugal remained strong, however, and its frequency became well established. Proof of its effectiveness came in the form of stories about listeners in eastern Europe. James E. Chase, director of the General Conference communication department at the time, wrote in *Tell* magazine in 1979: "A certain Hungarian man heard the *Voice of Hope* on his transistor set, and truth went to his heart. He wrote Paris for the Bible-correspondence lessons and went on to complete the course. Then there was silence.

"For two years," Chase wrote, "the Bible school did not hear from him. But the truth that had been transmitted into his heart was taking root, resulting in changes in his lifestyle and his relationship to Jesus Christ. As his life touched others during that two years of silence, he shared with them his wonderful newfound faith through personal testimony and Bible studies.

"Finally, the period of silence was broken by his request for the speaker of the Hungarian *Voice of Hope* program to come and baptize him and others. They all wished to do the Lord's will completely. Fortunately, the speaker was able to respond to the request. He examined their spiritual understanding and was so fully persuaded of their experience that they were not only baptized, but soon the group was organized into a small church."

The seeds sown by radio are hard to trace once they've sprouted. But the important thing is that the seeds are sown. A very interesting aspect of the radio ministry that we discovered over the years was that usually, when a radio listener finally asks for baptism, it is an unemotional decision based on many months, even years, of listening to our broadcasts. Well before the listener asks for baptism, he or she has identified with the teachings of the church and is fully acquainted with Christian beliefs.

Heinz Hopf, communication director of the Euro-Africa Division, sent in another story in 1982: "The place is a remote corner, a village not easy to reach, somewhere in eastern Europe. The scene: Every Sunday morning a group of about ten people meet in the home of family C, who possesses a radio set, a rather old but good shortwave receiver. There they listen together to the *Voice of Hope* from faraway Portugal. Months pass. Finally, the pastor of an Adventist church decides to begin evangelistic activities in that area of his district. He mails invitations to all neighboring villages. By 'chance,' an invitation comes into the home of that family. Once they attend the first meeting and find out that it is the same message they have listened to on radio, they do not miss a meeting. Through winter's ice and snow they ride bicycles the ten miles from their village to the meeting place. Today they and two of their neighbors are preparing for baptism."

The year 1978 turned out to be a milestone in the history of AWR. General Conference leadership spelled out its vision of encircling the globe with the Adventist message. In the spring of the year, the presidential administrative committee decided to ask the church in its annual council to approve moving ahead "with a denomination-ally owned radio station in Liberia, Africa, as funding becomes available and a continuing study of programming by the Far Eastern Division for broadcasts to mainland China and acquiring a new station in Guatemala."

It was the first time Guatemala had come into the picture, but I soon found out why. Ron wrote on May 8, "Guatemala is moving right ahead on constructing AM, FM, and shortwave stations. I've talked by ham radio to Bob Folkenberg (even played excerpts of the conversation on our World DX News program here in Europe). While not at this time strictly an AWR project, Bob hopes AWR will eventually adopt it." Two years later, on March 28, 1980, AWR-Latin America was born when shortwave broadcasts began from Guatemala.

It was also the first time Bob Folkenberg came into the AWR picture, but it certainly would not be the last. President of the church's Central American Union and son of Barbara and Stanley Folkenberg, he quickly realized that radio waves could reach far into the jungles and mountains of his mission's territory: Costa Rica, Belize, Guatemala, El Salvador, Honduras, Nicaragua, and Panama.

A man of many talents, Bob was most of all a consummate communicator, quick to grasp any method available to spread the Word of God. His concept of how to reach the world included use, not only of radio transmitters, but mail, telephones, television, computers, and any other technology at hand. His abilities attracted the church's attention sufficiently that in 1990 he was elected president of the General Conference.

As for the effectiveness of radio, he told an experience that convinced him of its universality. Riding a donkey one day through drizzling rain in the forests of Central America, he heard the sound of music wafting down the hillside. Peering out from under his poncho, he saw another traveler astride a donkey coming down the muddy mountain path, carrying a radio. It was the sound of HCJB, the famous missionary radio "Voice of the Andes." He decided then and there what the Adventist Church must do, and he went on to found AWR-Latin America, later renamed AWR-Pan America.

AWR was moving rapidly during the early 1980s, and momentum was building for a major event in 1985. Neal Wilson, elected General Conference president in 1980, encouraged church broadcasters by declaring, in his address at the quinquennial business meeting: "In my judgment, the use of the electronic media must be expanded beyond anything we know at the present time."

In 1982, Mike Wiist was called to manage AWR-Europe when Ron Myers left to manage two stations of his own in

France. A decision was made to build a small shortwave station at Forli, in northern Italy, in expectation that the government there would soon make broadcasting legal.

Daniel Grisier was appointed manager of AWR operations in Africa in late 1983. The General Conference agreed to sponsor an hour a day in French and English on Africa's most powerful radio station, Radio Africa #1, in Gabon.

My graduation from Andrews University was nearing, and I was anxious to return to the world of international broadcasting. Would I be needed in the new AWR? I hoped so, because it was the most exciting prospect I could imagine.

Chapter 11

Asia Calls

"This is the day I've been working toward, but I sure do wish it was already over," I confided to my wife. It was the day of the big test, the oral defense of my doctoral dissertation. I was confident in my preparation—many people told me, "If anyone knows this subject, you should, after all the months of study and research." It's what they tell everyone who has the preexam jitters.

"A Model for Development of a Telecommunications Satellite Network for Administrative, Educational, and Other Purposes in a Private Organization," read the front page of my defense program. The name was so long I could barely remember it, but it looked impressive, and I hoped the technological aspects of the topic would intimidate the examining committee enough that they would take it easy on the questioning.

The examining session lasted several hours. The six professors grilled me one after the other. I found myself actually enjoying the opportunity to defend my topic, one that I fully believed in. It was a formula for the church for establishing a satellite network to provide distribution of radio/TV signals to broadcast stations and for providing an educational TV channel for use in churches, schools, and hospitals. If nothing else, I hoped the university would be proud to have it lined up along with all the other dissertations in the James White Library.

A number of friends had requested permission to attend my defense, and before I knew it, they were hugging me and patting me on the back as the dean introduced me as "Dr. Steele."

With diploma in hand, I was now ready to go out and serve my church in a broader field again. It was time to contact the new communication director at the General Conference, Bob Nixon. I told him I was available to AWR again and hoped there might be a place where I could serve in some way. He said he would investigate the possibilities and let me know.

"Would you be interested in heading up our new radio-station project for Asia?" he asked a few weeks later. Would I? And how! It was the answer to my prayers.

"Thanks to the good help of our good friend, engineer George Jacobs, it looks like we'll soon receive a license from the federal government for this station, and we are anxious to start building," he continued, but I hardly heard what he was saying. My mind was already racing through the work ahead and planning our move to a tropical island in the western Pacific.

Most people begin their Guam stories by talking about the weather. "The warm, humid night air hit me full in the face," or something like that. Coming from snow country in midwinter to the tropical island's eternal summer was a delight to me. I love warm weather, so I prefer to start my story about Guam by telling of the people.

"Welcome to paradise," said the mission president's wife, Linda Bauer. Her husband, Bruce, was away on a trip, so she had come to welcome us with flowers at the airport at four in the morning! Obviously the air schedules in the mid-1980s were organized to serve departure cities, not the arrival city of Agaña, Guam. But how could you please everyone when it was a seven-hour flight from Honolulu and you crossed the International Date Line? We had lost a day and were arriving in the middle of the night.

Everyone welcomed us so warmly we couldn't help but know we were wanted. We soon discovered that Guam is one of the most hospitable places on earth. Any excuse is used for a "fiesta." Weddings, birthdays, anniversaries, baptisms, new building, new office, new home, anything serves as an occasion. A rule of thumb is to prepare twice as much food as you need so that all will marvel at the bounty and there will be some left over for those who stay longer and need another meal later in the day.

One of the first things we wanted to do once we were settled in our temporary home was to visit the property where the station would be built. Tulio Haylock, associate director of the General Conference communication department, had spent many months traveling to the Philippines, Korea, Palau, and other islands of the Pacific searching for an appropriate site. Finally, all the factors needed to build a radio station came together on Guam.

A most positive aspect of Guam is that it is a United States territory, subject to U.S. licensing and regulations. That meant we had total freedom to apply for a license, and once we received it, we knew it would be protected by U.S. law.

"And the land was without form and void," I paraphrased Scripture when I saw the property. It was in the hills just south of the town of Agat on the southwest coast, overlooking the blue Pacific. The site was on the extended slope of Mt. Lamlam, the tallest mountain (about 1,300 feet high) on Guam. Scant vegetation covered the area, and large patches of eroded soil marred the terrain, making one section look like a mini Grand Canyon.

"How will we ever erect large curtain antennas on this hilly place?" I murmured under my breath. It was obvious that some ingenious planning would be needed if AWR was going to be successful here.

Ray James, Guam-Micronesia Mission president, believed it was a miracle when his eyes saw the "property for sale" column in the *Pacific Daily News*. It was a classified ad offering twenty-two acres for sale. Large parcels of land were scarce on the island, but here was a plot just where we needed it. An option to buy was quickly arranged, and now all we needed was a building permit from the local government.

This would be the critical piece of our puzzle. Before we could be issued a permit, we would have to submit a building plan for approval, obtain permission of all government agencies, and hold a public hearing. And much of this depended on goodwill and good presentations of our project. Our work was cut out for us.

The project also presented a planning challenge. The General Conference committee, inspired by president Neal Wilson, had agreed to permit this station to be the recipient of the official offering that would be taken at the GC session in New Orleans in 1985. We had to see if the entire project, costing an estimated five million dollars, would fit into the funds that would be raised.

Fortunately, God's timing was to prove perfect, because the offering was taken just as we began to build the station, and we eventually had the privilege of dedicating the facility debt free. In AWR's planning, we decided to build it to accommodate four 100-kilowatt transmitters, even if two would have to be installed later. Four would be needed to cover Asia adequately. To care for this contingency, we would install four antennas, the largest being 330 feet high, a rectangle the size of a football field.

An initial step was to tromp through the field to determine exactly where the property boundaries were located. This was a challenging exercise, because the saw grass was high, the sun was hot, and the land was very uneven. One could easily fall into a hole or ditch with just a slight misstep.

"Do you see what I see?" I said to my colleagues after tromping half the day, measuring the distance from one palm tree to another, cutting my arms with saw grass, and sweating up a storm. "We don't have access to our land! We'll either have to build a bridge over the river and build a long road to the highway or buy some adjoining land so we can have direct land access to our property." In the end, we decided it would be cheaper to purchase more land, because we also needed more to accommodate the two antennas for northern Asia.

Three lots of about six acres each, owned by three sisters, abutted the twenty-two-acre lot we had already agreed to buy. The large lot already secured for us was owned by the three sisters' uncle. We would not only have to find a place in our budget for the additional property, but I would have to dip into family politics to convince the sisters to sell their land. This would call for some delicate diplomacy.

The Torreses were a charming and interesting family. The sons and daughters of the family patriarch had inherited the land at Facpi Point. It was subdivided into thirty-acre plots so that each son and daughter would receive an equal amount of inheritance. His sons and daughters, in turn, subdivided their plots into equal amounts for their children. The additional parcels we would need to build the station belonged to three granddaughters.

I wasted no time in making contact with the sisters. Two were willing to sell their land, but the one who owned the middle lot was hesitant. She was thinking of her son and wanted to have some land she could will to him. My negotiations with her stretched over several months.

A busy businesswoman, the only time she could see me was during her breakfast hour at Linda's Cafe, on the beach at Agana Bay. Over a large glass of orange juice, we discussed the possibilities. Reluctantly, she agreed to sell most of her lot, if a small plot large enough to build a house on could be reserved for her son. It was the only agreement we could make that would satisfy both parties.

With the land issue resolved, the next step was to petition the local government for permission to build. First, our plans had to be approved by the Land Management Department so that the property could be rezoned from farmland to light industrial. This involved several sittings with the Land Management Commission at their weekly meetings.

Finally, a public hearing was scheduled for June 19, 1985, so that citizens could give their opinions about our project. If we were to be successful in getting this final clearance for the project, we knew we would need to do a good bit of public relations.

"I'm glad I have a public-relations expert on our AWR staff," I said to Andrea as I opened our planning meeting. Our AWR board chairman, Lowell Bock, had made sure Andrea was voted PR and programming director. We appreciated his perceptiveness in realizing that her valuable talents would be necessary to the project.

"Well, since we are the only two people on this committee, we should be able to come up with a plan," she answered when I outlined the challenge before us. "First, we need to let all our church members on Guam know about our project. Some of them have contacts in government, and this will be helpful to us," she added. We immediately started sharing the details of our project with the church members at weekend meetings.

"We also need to try to affect public opinion. Maybe some newspaper advertisements will be the best way to do that." I could tell she was sharing an outline of a plan already developing in her mind. We immediately started planning four half-page ads for the island newspaper, the *Pacific Daily News*, owned by the Gannett Company. One would tie us in with our large medical clinic on the island, another would identify us with the Guam Adventist

Academy, a third would link us with the vegetarian restaurant operated by the mission, and the fourth would clearly show we were associated with the Adventist churches.

We had everything we needed for the public hearing, including a full endorsement by Guam's governor, Richard Bordallo, when, just three days before the event, I received a call from church headquarters saying we must cancel the public hearing lest the landowners decide to raise prices on the land we had negotiated to buy.

I decided this called for crisis management. I felt I knew the landowners well enough now that I could count on them to honor our agreements, and I knew that if the public hearing was canceled, we would be set back many months on our project. These two facts made me decide to ignore the order. I never heard any more about it.

The municipal building in Agat is a low, one-story building located right at the water's edge in the heart of town. Agat is known for its scenic beauty, its backwoods "boonie" areas, its water sports, and the cockfights down at the cockpit near Nimitz Beach.

In the municipal building is the mayor's office, next to the meeting room where our public hearing was to take place. Mayor Tony Babauta, we knew, was 100 percent behind our project. He became a great friend of AWR and was always asking how he could help. His support was crucial to our success.

At the appointed hour, officials of the Territorial Planning Commission arrived to join the mayor for the hearing. I was permitted to present our project using maps and graphs to indicate the type of radio station we planned to build.

The officials discussed the plan among themselves, then turned to the twenty or so people who had come to participate in the hearing. Thankfully, we had invited our church members, and they filled most of the seats. To our surprise, the land commissioner asked each person present to state whether he or she was for or against the project. The project was overwhelmingly approved by the group. A U.S. Navy representative was present, however, and she brought up some concerns about possible technical interference between our station and the naval operations at the nearby naval air station. We explained that navy permission had already been secured in Washington, but her complaint delayed approval by the Territorial Planning Commission by several months.

Another objection to our station came from another shortwave broadcaster that operated a station at the southern tip of the island. They were not represented at the public hearing, but their lawyer had written a formal objection to the Federal Communication Commission. Fortunately, the letter arrived two days after the commission granted AWR the license.

The United States government requires all radio stations to have a call sign, a four-letter identification. Traditionally, although there were a few exceptions, stations located east of the Mississippi River have a call sign that starts with the letter *w*. Those west of the Mississippi start with the letter *k*.

We weren't sure which side of the Mississippi we were located on, but we were told to choose a call sign starting with *k*, like all the other stations on Guam. After much deliberation, the call sign KSDA was chosen and approved.

Nearly everything was in place but the station itself. That was our big challenge.

Chapter 12

Rain Dance

"He says we can't make our road through his banana trees," Don Myers, our assistant chief engineer, said in dismay as I walked up to the bulldozer. We were finally ready to start grading a road to our AWR building site. We thought we could just follow the map made by the surveyor, but a nearby farmer discovered that the route went right through his small banana plantation, and he was not about to let us mow down his trees.

"Let's talk to him together and see if we can't work something out," I said as we turned to where the farmer was standing like a sentry at the edge of his garden.

"You can see here on the map where the access road is to be built," I tried to explain. "It's the only place the government will let us grade for a road," I added.

"That's OK, but it will take down a whole row of my best banana trees, so it's not OK, you see," he rejoined, quite agitated. We began to negotiate, and I soon decided that it wouldn't be neighborly for us to damage his crops like that, legally or otherwise.

"All right, we'll leave the row of banana trees, but we'll put the road in just to the right of your line of banana trees and hope we're not too far off the mark for this entry road," I suggested.

"That's OK; I'll go for that." He smiled. I knew we had made another friend for AWR. That was part of our mission as new citizens on the island. Our crew spent the next two hours digging up pepper plants and small banana trees to transplant them to an area our new friend indicated behind his house. Several months later, his wife, Maria, came to work for us as a custodian. She quickly became part of the AWR family.

We were so grateful to have the building permits and happy that we could finally go into action. But by now, in mid-September, we were well into the rainy season, and our greatest enemy would be the tropical rainstorms. Soon all our work clothes were the color of the red mud that covered our property.

We hadn't gotten more than two hundred feet into grading our new road when another delay loomed. A small hill, maybe forty feet high, was in the middle of our right of way. The Territorial Planning Commission declared we would have to have archaeologists from the University of Guam investigate the hill to see if it was of "archaeological value."

A number of new hotel sites on Tumon Bay, up in the center of the island, had found ancient Chamorro burial grounds on their construction sites, and the government had halted all construction until the ancient remains were removed. The process usually took six months to a year, and it became a headache to developers, who were required to pay for the archaeology work.

"Could this mound in our roadway actually be an old burial mound?" I anxiously asked the archaeologist when he arrived.

"Well, it's possible, but more than likely, it is just something left over from the war. We will know after just a couple hours of digging," he assured me.

Later in the day, he came to report. "There are definitely no ancient remains in that hill. I would guess it was built as a machine-gun emplacement by the Japanese during World War II." His words brought me great relief. Another potential delay to our project had been averted.

We were thankful for a dedicated crew of workers: Robert and Clarisse Sweede, retiree volunteers; engineers Butch McBride, Don Myers, and Brook Powers; and student missionaries Craig Caster, Gerald Kovalski, and Alan Carlson. Soon an islander from Truk, Domingko Saladier, joined our core team.

We felt overwhelmed by the task ahead, but everyone bent to the work, much of which involved sliding around in the muck between rain showers. As soon as the coral road (crushed coral is used in place of gravel in the making of roads on Guam) was completed, we began to hire work crews from the island labor force to prepare for concrete pours, assemble the six towers, then erect them with the four antennas hung between them.

The mission president invited us to use his office and another vacant office at the mission headquarters in Agaña Heights during the summer vacation months. But when September came, we were faced with the need to find a place for our headquarters.

"Excuse me, would you happen to know of anyone who wants to sell or rent their house in this area?" I asked at a house in southern Agat. I had decided to take my search directly to the people in the neighborhood closest to our construction site. At the third house, I found some encouragement.

"Well, I've heard that the young woman across the road there wants to sell her house." The man pointed across the street at a little stuccoed pink house. "She's just started working for Congressman Ben Blaz in Washington and needs the money to buy a house in D.C." As a U.S. territory, Guam is permitted to have a nonvoting member in Congress. Islanders are also permitted to elect the island governor and legislature but, by federal law, they cannot vote for president of the United States.

I thanked the man profusely for the information and hurried to call the number he had scribbled on a piece of paper. Soon we were negotiating for the house, which was located on the next-to-the-last subdivision lot nearest the radio station site, just a mile away. With AWR board approval, we soon became owners of what forever would be called "the pink house."

Over the course of the nearly two years it served as our office, it was broken into by thieves three times. A stray dog we named Lady came to be our official AWR watchdog until one night she disappeared, probably stolen by thieves, we guessed.

Occasionally a huge sow came to runt around the house in the heat of the day, and several times a hefty water buffalo came to peer in the windows to investigate our activities. Mayor Babauta told us that there was a herd of about two hundred water buffalo in the mountains above Agat that served as the source for new buffalo, or "tractors," for the farmers.

Worldwide interest in our project was incredible. We received letters from church members all around the world wishing us success. Many people telephoned to see how we were progressing. We decided to start a newsletter, *AWR-Asia News*, to keep people informed.

In some ways, the unusual interest in the "Radio Guam Project," as it was called by many, made our work difficult. There were people so taken up with the excitement of the project that they felt possessive of it and us. One man, who expected that he would be asked to head up AWR-Asia, went so far as to call all the public utility agencies on Guam to inform them of our needs and secure information, ostensibly for the project. I discovered this when I made initial contacts with the power company, water company, and telephone company. "Oh yes, someone called us on this, and we sent them the information," they said.

But at the same time, praise the Lord, thousands of people were solidly backing the new station with their finances. We were delighted to hear that the General Conference offering total was close to four million dollars. We still needed another million dollars to finish the station, and we were happy to hear General Conference treasurer Don Gilbert and his associate, Don Robinson, predict that that amount would come in during the next year as donations.

To add to the excitement, we were informed that Neal Wilson himself was heading up a campaign for an endowment fund, the interest from which would provide an annual budget to operate the station. Within a year, four million dollars had been pledged for this special fund. By 1992 the endowment topped $5.5 million, and the interest did, indeed, provide a substantial part of the AWR-Asia yearly budget. We were humbled to be involved in such an important activity for the church and so grateful for the grand support. How important it is, we decided, for missionaries to know that the folk back home are solidly behind them, praying and paying so the Lord's work can be advanced to the farthest limits of earth.

Another great challenge we faced was the preparation of programs to put on the air. This required travel by Andrea and me around Asia to help prepare the producers for the great task ahead. "This station will be like a giant elephant that can never get enough grass to eat," we said wherever we went, trying to put the task ahead in graphic terms.

Some studios already existed around Asia: in Japan, Korea, the Philippines, Indonesia, and Thailand. But many countries would have to start from scratch: Burma, Bangladesh, Sri Lanka, Papua-New Guinea, and Nepal. By the end of 1995, all these countries had new studios actively making programs, thanks, largely, to financial help from AWR supporters.

The Adventist Communication Center in Poona (Pune), India, would need to gear up for increased production, and within a few years, it became the center producing programs in the largest number of languages. China was our greatest target, and the General Conference took actions to see that a large new production center was organized in Hong Kong for this important work. Veteran China missionary Carl Currie was asked to move to Hong Kong to get this project underway.

There was also a need to find some way to prepare programs for areas such as Vietnam, Cambodia, and Laos, where we were not yet permitted to produce programs within the country. While we were able to bring the importance of these areas to the attention of the church, it was not until Gordon Retzer became manager of the station in the early 1990s that headway was made with the Vietnamese programs.

Meanwhile, back at the Facpi Point construction site, rainfall was keeping progress at a snail's pace. Guam's average annual rainfall is 85.41 inches. During June of 1985, we received a record 13.36 inches of rain—8.32 inches more than normal. This was to be the trend during our two years of construction.

"The rain just won't stop," our chief engineer complained. "It will take us forever to get this done at this rate." The problem of rain became part of our daily conversation with the Lord. We just had to keep telling ourselves that this was His project, and He would see to it that it got done.

Nevertheless, we had to keep moving ahead. Probably our lowest point came one day late in 1986 when we had a special cement pour at tower base number 1. Nine truckloads of cement were on order for an early-morning delivery. The trucks were late in arriving, and at midday, a steady downpour commenced.

All afternoon, the lead truck attempted to drive up the hill to the pour site but could not make it through the mud. The AWR bulldozer tried to pull it up, but without success. Finally, the engineers tried to carry the wet cement from the truck to the base site at the top of the hill and were able to fill some anchor holes, but the process was not fast enough to guarantee even half of the cement would make it to the top before it hardened in the mixer trucks. Finally, much of the cement had to be poured out on the ground, and delivery was rescheduled for another day.

The project was truly international. For our building, a Filipino architect presented the lowest bid. Our surveyor was Thai, and a Korean company won the bid for the building's construction. Meanwhile, most of the laborers on the project were Guamanians or other Micronesian islanders.

Every day had its victories. We rejoiced to see the foundation of our nine-thousand-square-foot building finished. It required 128 cubic yards of cement brought by sixteen large cement trucks. The walls seemed to go up quickly, almost overnight. The roof pour, on August 19, 1986, was thrilling, as 350 tons of concrete were spewed out of a cement pump truck that looked like a giant elephant with an extra-long trunk.

The antennas included thousands of parts that had to be assembled by hand. There were fifty galvanized steel anchor rods and twenty-six thousand pounds of guy wire cable. The towers came to Guam in four forty-foot containers. All together, the two giant tower/antenna systems arrived in more than forty thousand pieces. While the concrete bases were being prepared at the tower sites, the towers were being constructed at a twelve thousand-square-foot warehouse rented by AWR on the north side of Agat.

The logistics of the project were staggering for our small crew. The days were long, hot, tiring, and always hampered by the tropical rains. Guam has two seasons: rainy and dry. By the end of 1985, we had passed through the rainy season and moved into the dry season, but the rains continued. We learned to just keep working, paying no heed to the rain.

Slowly things took shape, and we began to have a vision of how our station would soon function. By mid-1986, our hopes were soaring, and we felt a growing optimism that would fully explode by the time of our scheduled inauguration on January 18, 1987. The date had been set, and all hands began to point toward that deadline. But in our minds, we were all asking ourselves, *Will we be ready?*

Chapter 13

The Trumpet Sounds

"Well, we're only a week away from inauguration, and we still don't have an antenna ready or the road to the building paved," our new treasurer, Marvin Baldwin, was sharing his thoughts out loud. A retired church treasurer, Marvin and his wife, Marie, had agreed to come help us finish up the construction project. They were mainly responsible for the modest yet attractive interior decor and furniture of the AWR building. Prior to their arrival, another retired church treasurer and his wife, William and Mona Pascoe, put in a stint as our AWR-Asia treasurer. I forever marveled at how the Lord provided the right people for our project at exactly the right time.

"Only the antenna will be a problem," I reassured Marvin. "The paving company has promised to have the road in by Friday. So at least we'll have the building and road ready."

The final week of preparation was an intense one. The General Conference president, two vice-presidents, the world church secretary, and other officials were already scheduled to arrive for the inaugural weekend, so the date could not be changed. Guam's new governor, Joseph Ada, was prepared to join the church leaders in the inauguration.

We invited the *Voice of Prophecy* choir from Korea to come sing for the ceremonies, and they raised their own plane fares for the trip. A brass ensemble from the Guam Seventh-day Adventist churches had been practicing for many months for the occasion. Several town mayors and senators were invited.

For the vegetarian fiesta meal, we asked the help of Mariquita Taitague, wife of local Pastor Frank Taitague (later elected president of the Guam-Micronesia Mission). Mariquita was known in the church as Guam "hostess par excellence." She and an army of church women were planning a lavish display of island cuisine for the event.

Two days before the big weekend, the road pavers showed up with their full complement of equipment. In two days, they had the new road, nearly a half mile in length, paved right up to the building. It wound from the main highway over "machine gun hill," down to the station property, through the former mini Grand Canyon, and then up to the AWR building. It was beautiful and finished off the site in a most spectacular way.

Finally, the great day arrived. Nearly five hundred people assembled in front of the gleaming white radio station building for the ceremony. In front of the building, we erected flags representing the twenty-two countries to which AWR-Asia would be broadcasting. The church and island officials arrived in a long motorcade escorted by Guam police.

Governor Ada, in his remarks at the ceremony, praised the Adventist Church for its accomplishments on Guam, especially in matters dealing with health and human life. He further wished AWR-Asia success in its "mission to bring a message of hope—a promise of a better tomorrow—to millions of people yearning to be free."

Neal Wilson said, "A long-cherished dream is fulfilled, and many fervent prayers answered." Andrea and I gave each other a knowing look. We knew it was not only the dream of all of us who had been called to work on the project, but that the president and church members around the world had been dreaming of this day for a long time, too, and they were extremely proud of the accomplishment the church had made in this new facility.

After the speeches, we took the governor on a tour of the station. We were proud of the large L-shaped building. It was bright and functional. The large transmitter room was designed to accommodate two more transmitters someday in the future. The entrance, featuring a long, permanent canopy for protection from rain, was situated at the center of the building. Many glass windows in the area of the studios and master control room made observation by visitors convenient. Skylights down the long hall of the administrative wing made the building bright even without lights. The roof was designed to serve as a catchment system to collect water in a large basin behind the building.

After the tour, Governor Ada rejoined Neal Wilson at the front of the building, and they went through the fiesta line to start the meal. We were anxious to expedite the governor's visit because we had been told that he could not stay long for our activity. We were surprised that he stayed more than an hour after the meal, just chatting with President Wilson.

Our staff was very exhausted after the big events surrounding the inauguration. We were disappointed that we were not yet on the air. But work continued on erection of antennas 1 and 2. One person who was determined the antennas would go up was a student missionary, Reinhold Grellmann. Even when our engineers were off the island on other business, Reinhold worked overtime to get the antenna up so we could go on the air.

He epitomized the army of young workers who came over the years, about twenty-five of them during our seven years on Guam, without whom we could not have survived. They worked in engineering, operations, programming, public relations, and other areas. Each would take a year out of college study to gain practical experience at AWR. Working conditions were sometimes appalling, always difficult and demanding, but they bore the situation well and contributed greatly to AWR-Asia.

Our staff members shouted with joy, hugged each other, and cried tears of happiness when test broadcasts began at 2:35 p.m., Guam time, on March 5, 1987. Calls made to Hong Kong and Singapore verified that the signal was indeed being received in those locations. Regular programming began at 7:00 p.m. on Friday, March 6, just as the sun set on the rim of the Philippine Sea. Church workers from the Guam-Micronesia Mission, the Guam SDA Clinic, and the Guam Adventist Academy witnessed this historic event and rejoiced with the staff.

The staff of the Korean *Voice of Prophecy* studio sent a telex reporting good reception there during test broadcasts. The first reception report by mail arrived the following Monday from a young listener in Japan. By the end of the month, more than one hundred letters had arrived from twenty different countries: Japan, Korea, Hong Kong, Singapore, China, Malaysia, Indonesia, the Philippines, Australia, New Zealand, Italy, and the United States.

We read each letter with heightening excitement:

"I'm very pleased in sending you this report. I've been waiting for this day for a long time. Congratulations on the birth of AWR-Asia."—Japan.

"A good, clear signal most of the time . . . very best wishes for great success with KSDA."—Australia.

"Looking forward to some interesting programs from you!"—New Zealand.

"What a thrill it was to hear the words 'this is KSDA, Adventist World Radio-Asia,' and to hear 'Lift Up the Trumpet.' This is what I have been praying and working for many years—to see the church proclaiming the gospel to the world with a LOUD VOICE!"—Texas.

"I am pleased to inform you that the AWR-Asia broadcast reaches me clearly in North Sumatra, Indonesia." —Indonesia.

"Thank you very much for your broadcast. I am not a Christian, but I enjoy your English program. I promise to continue listening to your broadcast."—Japan.

In the same newsletter in which we shared the first listener letters with our supporters, we had to announce the end of our publication. A decision had been made at church headquarters to discontinue our newsletter in favor of a new AWR worldwide newsletter to keep supporters informed.

For a year we were without *AWR-Asia News*, and donations began to drop drastically. By the end of the year, an appeal from the GC treasury to do something to bring back the financial support resulted in orders for our newsletter to reappear. It did, in the summer of 1988, with a new computerized format. Soon our donations picked back up, and we were able to recapture much of the interest that had been generated early on in the Guam project.

An early success story for AWR in Asia was the country of Burma (now Myanmar). In response to the broadcasts, our studio in Rangoon (now Yangon) began receiving scores of letters from most of the seven states and seven municipal divisions of the country. The communication director wrote: "Listener letters are coming from areas where we have never entered and have no church members," and "Requests for the Bible course comprise almost half of all mail received!" This was, indeed, a good report, because our average request for Bible courses from the rest of the world ranged between 10 and 20 percent.

One man wrote that he had organized a group of eleven people to listen to the broadcasts every day. The group was his family. "My nine children are 22, 20, 18, 16, 14,

12, 10, 8 and the youngest is 6 years old," wrote the forty-three-year-old businessman. He closed his letter by asking AWR to please enroll "us" in the Bible-correspondence course. ⊕

A teenager wrote to say, "I am not a Christian, but after listening to your broadcast, I enrolled in the Bible-correspondence course." He wanted to report that he had found someone to study the Bible with him: his brother, a Buddhist monk! ⊕

Chin Chu, a police officer, was baptized in early 1989 after hearing the Adventist message first on AWR-Asia. During a period of disappointment in his life, Mr. Chin heard the station theme song, "Lift Up the Trumpet," on his portable radio. The song made him forget his anger; he wrote for the Bible-correspondence course, completed the studies, and was baptized. ⊕

The Burma Union, acting to take advantage of the new interest shown in the Adventist message, did its best to prepare people to go into these new areas. We were told of a young man, a former national wrestling champion, who had become a church member. He was not gifted as a public speaker, but he was sent to a mountainous area where our church had never worked before.

His only tools for his work were a Bible, his guitar, and a shortwave radio. His method was to appear in the town square, start singing songs with the children and others who gathered, then read some Bible texts. Then, to complete the program, he would turn on the radio and invite everyone to listen to AWR. His efforts were very successful, and a number of new church groups developed in that area. Other young people, usually recent graduates of the theological seminary, were sent to areas where new groups were forming as a result of the radio broadcasts. Dozens of new churches began in this way.

At the invitation of listeners in a village where there were no Christians, members of the radio production team from Rangoon held an evangelistic meeting in May of 1990. Four people were baptized. ⊕

In March of 1991, the production team again held meetings, this time in the capital city of one of the states of Myanmar. Seven people were baptized.

Probably the most incredible story to reach us from Myanmar was that of May Lwin, an eighteen-year-old girl who lived in a remote section of the country. She heard our programs on her father's battered shortwave receiver and decided to enroll in the Bible-correspondence course. She decided to be a Christian, the only one in a village of spirit worshipers.

But "the Evil One" was angry. A villager, jealous of May Lwin's happiness, one day appeared at her family home to announce an evil curse on the family. May Lwin's parents demanded that she renounce Christianity lest they all suffer a great calamity. But she refused. They then said she must leave the family. Desperate in her desire to follow her new faith, she sought the only way out that would keep her faith intact and maintain respect for her parents. It was an acceptable act among her superstitious people to take poison.

Her family prepared to mourn her departure, but she did not die. A second time she took poison, but to everyone's amazement, she didn't die. Such a thing had never happened before in her village!

"She serves a powerful God. He did not let her die after taking the poison," her father said in stunned amazement. "I must let her worship as she wishes." His judgment was accepted by the villagers. Today, a new Adventist church in her village is a witness to the power of the true God. ⊕

Chapter 14

To All of Asia

"The view is spectacular," Brook Powers, our assistant engineer, said as he landed next to the tower he had just been to the top of. "You can see all the southwest coast, from Piti Power Plant down to Cocos Island." His arm waved in a large arc from north to south. My mind flashed back to when I had met Brook three years before. He was an engineering student at Andrews University, and we had needed engineering help at WAUS. He turned out to be an excellent worker, and I sensed the potential this sensible, hard-working, bright young man might have for the church.

"Why don't you come work for us on Guam?" I asked him point-blank one day after I decided to take the job on Guam myself.

"Yeah, why not?" He beamed, as if we were swapping jokes.

"I'm serious, Brook," I said. "We have a big job to do there, and we'll need your help." Apparently he had never seen a future in which he would drift very far from his home in the American Midwest.

"I'll be glad to come help out for a couple years," he told me several weeks later. I accepted his offer and arranged to hire him. He was one of the first to arrive on

Guam to help build the new station. Two years later, he invited his fiancée, Patricia, to come visit Guam. Soon they married and returned to us as a couple. After a short while together on Guam, they made another decision about missionary service.

"We've discussed our future, and we've decided that this is where the Lord wants us to be," Brook reported. I couldn't have been more delighted. As it turned out, he became a mainstay for our radio ministry to Asia, serving under chief engineers Butch McBride, Don Myers, and Elvin Vence.

Some months later, we were able to offer a job as program director to a young man who wrote from France, where he had been working for Ron Myers at his station in northern Italy. Greg Scott became a valuable asset to our team and several years later married a young woman, May Lazo, who had come as a student volunteer for one year.

We all fell in love with Guam. Soon after arriving, I became involved with Pathfinders again, devoting many weekends to camping in the jungles or at the beach. Together we climbed the mountains; stood under waterfalls that plunged from the cliffs; swam in underground pools housed in dark, cool caves; and hiked the white sandy beaches.

But one of the biggest challenges our staff faced on the island came as we realized we must build a new church in Agat. Some years before our arrival, a Filipino member in the church, Jelly Macadagum, had formed a company in the village by hosting Bible study and church services on the porch of her house in Santa Rita.

The small group grew to about fifty members, enough to organize a full-fledged church. A pastor, Ely Jimeno, was assigned to the group, and the congregation began to rent a Baptist church in the heart of Agat.

"We have purchased a lot just off the main highway," Ely came to announce one day in May 1986. "Now we must get to work and build our church." He looked hopefully at our staff, anticipating our help. We had some experience in construction, after all, so we were eager to help build the new church.

The next two years kept us busy raising funds and spending all-day Sundays at the construction site. First we worked in the blazing sun preparing the foundation. Then we worked in the blazing sun putting up the cement block walls. Then we worked in the blazing sun putting on the roof. Of course it was hard work, but the church grew tremendously as a spiritual family.

It was with great joy that we occupied the new building, the fourth Adventist church on Guam and the only one with air conditioning. It was valued at a quarter million dollars and was debt free when we started using it.

Meanwhile, the AWR station continued to reach out to all of Asia with the Adventist message, helping to start new church groups in many places. Letters indicated our audience was dispersed widely and growing. By 1989, letters had come in from 102 countries.

A man walked into the Kamenokoyama Church in Japan, where an evangelistic meeting was beginning, and registered with the receptionist. When asked how he learned about the meeting, he said, "By radio." When questioned further, the man said, "Yes, it was by radio. I am a faithful listener of the AWR–Asia Japanese program." ⊕

A teacher in the Philippines wrote to say how useful the AWR broadcasts are to her. She uses the things she learns from the radio programs in her classes. ⊕

In Indonesia, a religious leader who especially appreciated the family-life programs in Indonesian urged her parishioners to listen to the Adventist broadcasts. She said that the programs were excellent and that she took thoughts from the programs and adapted them to her work as a family counselor and church leader. ⊕

An Adventist pastor in the North Solomons Mission of

Papua New Guinea wrote to say he personally knew of another church in his village that every Sunday heard the Adventist message through their pastor, an AWR listener. The good man listened intently to AWR, then made up his sermon notes based on the broadcasts. ⊕

From north Sumatra in Indonesia, a listener wrote, "There are four families of us that always listen to your station, and we hope to promote it and invite our friends to listen. We would like a program schedule." ⊕

An Adventist physician making visits to remote villages in the Philippines reported that in several places groups of people gathered to listen to AWR-Asia on what sometimes was the only radio in the village. In one village, a man hooked up an amplifier to his radio so the rest of the village could hear better!

A letter from our program producer in Cebu, Philippines, confirmed the report. "Last week a couple from northern Cebu came to my office bringing their 200 pesos donation for the radio ministry. They said that in their village they are the only ones who have a shortwave receiver, but they are sharing it with their neighbors by connecting it to a powerful amplifier. As a result of this sharing, a company of believers are now meeting in their home." ⊕

A Hindu man in India with a family of five said that "through the Tamil programs my family has come to know Jesus as the true, living God. We realize that idol worship is sinful, and we have repented of all our sins. . . . We hope to be baptized soon!" ⊕

An evangelist in a Pentecostal church in Orissa said, "I have a congregation of 50 people. To feed them spiritually, I listen to the AWR Telugu broadcast and take notes, then share the message with my people on Sunday." ⊕

An Indian young man of twenty wrote, "The AWR Telugu programs help me make the right decisions. I encourage my friends to sit with me to listen to AWR, and at least a dozen of us enjoy the programs regularly." ⊕

A young Korean couple living in China arranged to visit relatives in South Korea. The husband's mother, who was a devoted AWR listener, encouraged them to make the trip and even helped them make their plans.

She insisted that they visit the AWR studio in Seoul so they could find out more about those people who were speaking on the radio. They had a delightful visit with our AWR studio team and agreed to begin taking Bible studies. They were soon baptized.

After several months in South Korea, the young couple returned to their homeland, accompanied by an Adventist pastor. Soon seven more people were baptized, and more were planning for baptism. ⊕

Joe Morgan was known by his friends as one who could "drink alcohol like a fish drinks water." As a supervisor at a giant copper mine in Papua New Guinea, he worked long days, then stopped at a bar for beer, and staggered home to his apprehensive family.

One day he staggered into his home and turned on his radio to hear the news. What he heard was the best news ever heard by any person—Adventist World Radio-Asia preaching the good news of salvation.

Listening to AWR became a habit, and then one Saturday morning, Joe announced to his family that they were all going to church! Soon they were enjoying the church meetings at the Rumba Adventist Church in Arawa. It wasn't long before the love of God's family convinced Joe to give up his liquor, betel nut, and cigarettes, and a baptismal day was set.

Only a few weeks before the planned baptismal day, a medical checkup revealed Joe had cancer and little time left to live. On his deathbed, Joe earnestly told his wife and six children, "I have no regret. I have made my choice to follow the Lord. Be faithful to the Lord, my dear ones, and I'll meet you in heaven." His wife and three teenage children were soon baptized, and at the funeral for Joe, many of his relatives indicated they would like to study the same Bible that had changed Joe's life so dramatically. ⊕

Agustinus listened intently to AWR day after day at his home in Indonesia. As he studied the *Voice of Prophecy* Bible lessons, he began to understand the commandments and more about Jesus' saving sacrifice. He said he felt he had been "far away from God." Then he met an Adventist pastor, who began to study the Bible with him.

"I told my younger brother about these studies, and he joined me," he reported. Both young men were baptized on April 29, 1990. "We faced difficulties with our family and relatives. I am the sixth and my brother is the seventh in our family of 12 brothers and sisters. Please pray for us here in our new faith," he pleaded. ⊕

Fighting God seemed to be a way of life for Takuya in Japan. He wrote, "I attended a Christian church when I was in high school. Then, during my college days, I forgot about God, and life didn't go so well. I even contemplated suicide. Only listening to the *Voice of Prophecy* on AWR kept me from doing that.

"Then I started attending church again, but often found myself arguing with the pastor about the beliefs I learned from AWR. I started searching for the true church and ended up visiting about 20 different Christian churches.

"Then one day God led me to the Seventh-day Adventist Church. I suddenly discovered I had found what I had been searching for and was baptized on May 26, 1990." ⊕

One evening, Mr. Maung, in Indochina, was tuning across the dial of his shortwave radio and was startled to hear a program that he had never heard before. The reception was excellent, and the program was in his native tongue, so he continued listening. Later that evening, he told his wife how much he had enjoyed the program and said they had announced another program in the morning.

Up bright and early as usual, the couple tuned their radio and sat listening. Through music and story and Scripture, the gospel message was opened to them. That night after supper, Mr. Maung gathered his children to listen. They especially liked the children's story.

So it became a habit—early in the morning and in the evening, month after month, with conviction of truth growing all the time. Finally, Mr. Maung wrote to the address given on the program, and the family began Bible studies. A year after that first day, he wrote again to the Bible school and declared that he and his whole family were ready to become Christians through baptism. He was assured that a pastor would come as soon as possible.

The soon became later, but when the pastor finally arrived, he found the family rejoicing in their newfound faith and firm in their conviction. "We told our friends about our intention to become Seventh-day Adventists," the father announced, "and some people came from another church and urged us to join them. But we told them we believed that the Adventists teach the Bible truth, and we would wait as long as necessary to join that church," he concluded triumphantly. ⊕

"I came to be baptized," Zentua announced to the *Voice of Prophecy* Bible School director in Jakarta, Indonesia. The young man, twenty years old, had come from north Sumatra to Indonesia's capital city. He was determined to find the people who had been sending him the Bible lessons he learned about on AWR.

He went on to explain that he had completed the *Voice of Prophecy* course and had been listening to AWR every afternoon. His Bible-school teacher began to question him about what he had learned and why he felt he was ready for baptism.

The teacher soon discovered that, indeed, Zentua was prepared for the important step into God's kingdom, but that a few weeks' further study would be valuable. Two months later, he was baptized and began his new Christian life as a seller of Christian literature.

Meanwhile, another young man arrived in Jakarta to enter the university. He knew no one in the city and was anxious to make friends. He decided to go to the only "friends" he knew about in the city—the people at the *Voice of Prophecy* Bible School, whose address he had been hearing for three months over AWR.

Surprisingly, he met a friend from his old hometown right there at the *Voice of Prophecy* school: Zentua. It was a joyous reunion that became a celebration shortly afterward when Marasati was baptized into the church—with Zentua at his side.

But this story has more than one good ending. Marasati was summoned one day by his family back in his village because his parents had become seriously ill. He returned to nurse them back to health, but also to share the story of his new, happy life. He encouraged everyone he met in the village to listen to the AWR broadcasts, which came in loud and clear on the radio.

It wasn't long before his family was ready to be baptized, too, to become the nucleus of a new church family in that town. Zentua went on to study theology at an Adventist college and is now a Seventh-day Adventist pastor. ⊕

Touching Hearts
in China

Our Guam station's primary target is mainland China, and results show that that target is being hit. In the late 1980s, it was estimated that there were between twenty and thirty thousand people joining the church each year in China. Of course, AWR wasn't the only influence leading to this success, but it played an important role. In the world's most populous country, free movement from province to province is restricted; open evangelism is not permitted in most areas; and where pastoral training was impossible, the radio beam from Guam provided a steady guide for the church.

During the first forty-five months that AWR-Asia was on the air, more than fifteen thousand letters arrived as a result of the Chinese broadcasts. In 1989 alone, AWR received nearly six thousand letters.

But it wasn't just letters we received. More than half of those who wrote in became students in the Bible-correspondence course. Based on the mail we received, we estimated that 80 percent of our listeners were young people.

Stories of groups of people, large and small, listening to the radio messages together came to our station constantly. University students seemed especially adept at forming groups to listen to the broadcasts. Weekend study classes were formed, and pleas for Bibles constantly came to the Bible School.

Adventist pastors were frequently surprised to find visitors in their congregations who had heard the message on radio and come in search of Sabbath-keeping believers. Usually these people had a full understanding of Adventist doctrines and were ready to become church members. Some listeners who accepted the Adventist message became self- or, we might say, God-appointed preachers.

Most of the Chinese listeners share their deep gratitude for the message of hope the AWR programs give.

Richard Liu, who succeeded Carl Currie as director of the Hong Kong studio, developed the concept of "College of the Air" and asked Southern College in the United States to send a religion professor to help with the on-air classes. Douglas Bennett, from the religion department, went to Hong Kong to record the programs.

In a remote western city of China, a young American woman sat listening to the college class on radio. "She had been sent by a Christian church to teach English at the economic school," wrote Douglas. "She began to listen to the broadcast, which was a class on preaching. Then she listened to classes on Revelation and Christian beliefs. She told me she got some Muslim people to listen with her, and now some of them are believers." His story also included the baptism of this young woman on November 26, 1994, on the campus of Southern College during a visit to the United States. ⊕

A twenty-one-year-old in Shanghai wrote: "Recently, through the radio waves created by God, I have been able to listen to your high-quality programs. They have helped me to know Jesus Christ, and now my heart is always filled with joy. I am going to do everything to follow Jesus with you. To help me, a seedling, to grow, please send me a Bible." ⊕

"I always listen to your broadcast. Your programs are like rain watering my dry heart, giving me strength to go on," wrote a listener who works in a chemical factory in Shandong. "Although I am physically sound, I am empty spiritually and in desperate need of spiritual support. Under such circumstances, your explanation of the Bible brings me to a land of great happiness." ⊕

From Guangdong came these words: "I am a faithful listener of your radio station, for I enjoy your programs very much. The speakers have become my good friends and teachers. You are the lighthouse that pushes me to go on, helping me to face reality and life with determination, helping me to believe firmly in God. I have chosen the brightest path in my life, the way of God."

Another listener wrote: "I am a regular listener. Not long ago I heard your program accidentally. In the past I did not have any religion, but after listening to your programs I gradually have begun to believe in God." He had to share some good news. "I recommend your programs to my friends, and many of them have come to like your programs and will become faithful believers, I'm sure." ⊕

University students revealed their search for meaning in life. A medical student wrote: "I was especially delighted to hear your program last night because it was my birthday! I plan to become a faithful listener. My classmates in the dormitory also enjoy your program and were the ones who told me about it." Another medical student said: "In my class, twelve out of twenty-four students listen to your program. Four of them are women." ⊕

From another city, a student wrote: "Every night I join my friends in the dormitory as we gather around the radio to listen to your program." Another student who had failed to pass the university entrance examination was ready to give up. He said, "My expectations turned into bubbles;

my life became so empty that I lost faith in myself. Yet while I was sinking into deep sorrow, I came into contact with your broadcast. Your voice was like a flowing stream, melting the ice in my heart and lighting a fire of hope. I continue to listen to your broadcast every day." ⊕

Another letter brought news of a small home church where twenty to thirty people met on the front lawn. The owner told of his personal revival ignited by the radio broadcast: "I learned of the Sabbath when I was 8, and now I am 57." There were only nine families in his small village. Three were Sunday keepers; five had no religion. "Other members of the home church have to walk a long way to worship with us. The closest person walks one-and-a-half hours, and one family walks four-and-a-half hours. We are very happy to hear your broadcast. We read the Scripture and sing hymns with you." ⊕

An Adventist Chinese American made a long-awaited and much-anticipated trip to his homeland to visit relatives in a large city of China. Though not Seventh-day Adventists, they were sure they knew where an Adventist church was and readily agreed to take him there on Sabbath.

Arriving at the church, they found the doors locked. A neighbor said that the church services were only on Sundays. Disappointed, they turned to leave but saw a large group of people approaching. Noticing that the group's leader had a Bible under his arm, they asked, "Are you Christians?"

"Yes, we're Seventh-day Adventist Christians," they replied. "Our church doesn't have a baptistery, so we have arranged to have our AWR baptism in this church."

The visiting American and his relatives joined the group inside and watched as nearly two hundred people were baptized—all as a result of the outreach of AWR. ⊕

The only sound heard in the broad expanse of rice fields ripe for the harvest was the soft serration of the scythe against the rice stalks. But a voice was being heard in heaven: While she was bent over in the rice paddy, swinging her scythe rhythmically through the ripened grain, Mrs. Wong was also praying constantly that someone would send a pastor to her village in Manchuria.

She and many other families in the village listened to shortwave radio to keep in touch with the world, much of it so different from their quiet farming town in the northeast of China. And the programs they listened to were all different too. News, sports, politics, and religion vied for attention.

But there was one program they kept going back to: it was in their native tongue, and the words of the speaker and the ideas he presented were new, exciting, and filled with hope. That program was being broadcast over AWR.

Then one day, the Lord answered Mrs. Wong's prayer. Into her village walked a friendly and interested man. He went straight to one home and after a time came out, accompanied by the resident, who took him to another home and then to another and finally to see Mrs. Wong.

"Hello," the man said. "I understand you have written to the Bible-correspondence school in my country. Many of your neighbors did too. They sent me to teach you more about Jesus. I am Pastor Lee."

And so, in one of the adobe and thatched-roof houses, connected to each other in long rows for coolness in the summer and warmth in the winter, the Adventist message was spoken out loud for the first time. Mrs. Wong joined the other joyful radio listeners for the first-ever Sabbath service.

During this first visit, Pastor Lee baptized seven people. Then he had to leave. Some months later, Pastor Lee returned and baptized seven more people. ⊕

In another city, an AWR listener, Mrs. Kwan, asked her son and daughter-in-law to take a letter to the Bible-correspondence school in Seoul when they visited South

Korea. The Kwans were not Christians but learned about Christian beliefs while staying with their Protestant relatives in Seoul.

When they visited the Bible School office, they had a chance to hear the Adventist message. They were given a tour of several Adventist institutions, and a pastor gave them Bible studies. Soon they were believers, members of the church. After their baptism, they received many gifts to take back to China—clothes, books, radios, and a tape recorder.

The same Pastor Lee who went to the other village accompanied the Kwans to China and began to teach the message to others in their village. As a result, the mother of the newly baptized couple and six other people became believers. These nine people then began sharing their faith with their neighbors. ⊕

Sitting next to her radio, pen poised to write down texts and thoughts that impressed her, Mrs. Bao was continuing a habit that began when she discovered the AWR broadcast. On this day, though she did not know it, she would hear something that would forever change the way she worshiped—she would hear about the Lord's true Sabbath.

Already a Christian, Mrs. Bao was astonished. She began to study her Bible and became more convinced. But she was still confused: Why did so many people go to church on Sunday if Saturday was the true Sabbath? She decided to talk to her pastor. He told her that the day "had been changed," and now Christians keep Sunday "because Jesus was resurrected on that day."

Mrs. Bao, however, was not convinced that what he said was right—after all, she had been studying the Bible. So she returned home to study—and to listen more to her radio.

Finally, convicted beyond doubt, she realized that she would have to keep Sabbath by herself, if no one else was willing to do it. So she began in her own home. But she became happier and happier as she continued to listen and, study and she shared her discoveries with neighbors.

Soon ten people were meeting in the Bao home each Saturday, studying God's Word and praying for Jesus' soon return. She wrote: "Please send us an Adventist pastor to baptize us; we are ready." ⊕

In 1990 the International News Service reported from China: "In recent months students have been converting to Christianity in large numbers. The rate at which they are now turning to Christ is literally a dormitory at a time."

At about the same time, we received a report that confirmed this news. It was Christmas Eve, and the Adventist congregation in a certain city had planned a special Christmas program. As the members began to sing Christmas carols, something unusual happened.

The doorway opened, and in came students—dozens of them! They were from a nearby university and had heard about the meeting. They had first heard the name of Jesus on AWR. They were thirsty for words of hope— they wanted to learn about Christ, the Saviour of the world.

The church elders put their heads together and decided to alter their program a little so that the young people would hear the complete story of Jesus. The simple service became an evangelistic meeting—and what an attentive audience! Many of the students expressed a desire to study more, and they continued to attend that church regularly. ⊕

One day, Richard Liu, director of the Chinese work, called to tell us of an unusual story that appeared in the newspaper in one large city.

"The paper reports that a huge traffic jam had occurred in the city last week," he said. "It was an inconvenience to the people of the city, and the city fathers were not happy with the situation.

"The reason for the traffic jam was that one thousand people were being baptized in the river! This was an Adventist baptism," he said, "but we didn't even know there was an Adventist church in that city!"

The story was typical of those we were hearing more and more. Richard investigated and a few weeks later came back with the rest of the story: "We found out that several pastors from another location traveled to the city when they heard there were so many people to be baptized. But the city officials would not permit them to hold the ceremony because they were from outside the city. So the pastors ordained a number of local elders, and it was these elders who baptized the people who had learned about Adventists by radio. But the most surprising aspect of the story is that the newspaper didn't want to emphasize the baptisms, so the article was very conservative on the number who were baptized. The elders there say it was closer to *two* thousand people who were baptized!" ⊕

Within eighteen months of commencement of the broadcasts to China from Guam, AWR had received letters from all the provinces of the great country. One area we still hoped would respond, a place that we all held in awe, was the Himalayan tabletop of Tibet. Finally, in late 1988, it came: "From your program I have learned some English," wrote the young woman. She worked at a weather observatory. "I enjoy your programs very much. It brings us closer. I dedicate this little poem to you and wish you a happy new year." Enclosed with her letter was another sheet with this handwritten poem:

From the clear blue sky, through the pure white clouds,
Slowly the melody reaches my ears.
I hear your voices from the far side of the ocean.
At the same hour, each day we come together.
The wave of sound brings blessings across the ocean.
Oh, radio waves, may you accompany us across the endless years.

Chapter 16

Pacific Blue

"Living on a small island, you're sort of a victim of the sea," our neighbor, Joe, said one day. He was a product of Guam, a Chamorro, and was always sharing bits of advice and wisdom with me from his store of folklore and island tradition.

The island is just a tiny volcanic speck, thirty miles long and from four to twelve miles wide in the midst of the mighty Pacific. It is, however, the largest island of Micronesia and serves as the financial and transportation center for the region.

Because of the mountains in the south, the island is sparsely populated. A number of streams, called "rivers" on Guam, spurt out of the mountain heads to fall and tumble down to the ocean. One of my favorite hikes follows the Sella River up the side of the southern hump of Mt. Lamlam, within view of our radio station. The trail leads to the river's source.

At every corner of the stream, water comes gushing through huge rocks to form small pools, some large enough to jump into and get instant relief from the stifling heat. Then the water hurries down another series of boulders to splash into the next pool. The river looks like a mountain stream that Walt Disney World would have spent millions of dollars to create. At the source, on top of the mountain, water comes seeping out of the earth under dense undergrowth. It is the most wonderful, sweet water I have ever tasted.

Underground aquifers in the northern half of Guam provide sweet drinking water for the island. Another special hike for our Pathfinders was to trek to one of three caves that we knew had underground pools. There the water was cool and clear, the only place in Guam's perpetual summer weather where I ever felt I might actually get chilled.

In the south, one could easily get lost in the tall saw grass and jungle terrain. Once, on a hike with our Pathfinders across the mountains, I noticed one lad was missing. Quickly I turned back to look for him, calling his name as I walked down the trail. "I'm here, I'm here." I heard a pitiful voice in the distance. He had taken a wrong turn and was struggling to find the main trail in the six-foot-high saw grass when he heard me calling.

One of the most famous stories to come out of the southern jungles climaxed in 1972. In the closing days of the American invasion to take back the island from the Japanese in the summer of 1944, many Japanese soldiers, including Shoichi Yokoi, hid in the jungle to avoid being captured. They refused to surrender; to do so would have been dishonorable. For some years, Yokoi lived with two other Japanese soldiers in a cave a couple miles from Talofofo, a village on the southeastern side of the island.

To survive, they ate coconut, taro, soursop, and freshwater shrimp and eels caught in the Talofofo River. They fashioned shrimp traps from bamboo, made rope from coconut fiber, and clothing from wild hibiscus.

After thirteen years, the two other soldiers left Yokoi and moved to another cave. Later, they died in the jungle. It is unsure whether they died from natural causes or from suicide. Yokoi dug another cave about seven feet underground in the middle of a thick bamboo forest. There he lived until 1972, twenty-eight years after going into hiding!

One night, two Chamorro men, Manuel De Garcia and his brother-in-law, Jesus Duenas, were out hunting in the jungle near their Talofofo home when they spotted a moving figure in the thick bushes. Guns in hand, they ran after the creature and knocked it down. They were surprised to see that their prey was a thin, wild-haired, and even wilder-clothed Japanese man.

"I knew he had been living in the jungle when I knocked him down," Duenas said. "He had clothes like the skin of a tree on his body. He was scared; he thought I was going to shoot him." They took him to village officials, who learned his amazing story through an interpreter. He was repatriated to his home country, where he was considered a hero.

Mr. Yokoi and his wife were guests of honor at a ceremony on Guam in 1985 to dedicate a replica of his cave in the Talofofo Falls Park. Yokoi, at seventy-one years of age, said he still considered Guam his second home.

The idea of islanders being "victims of the sea" seems especially true when typhoons roar across the western Pacific each year. They usually start southeast of Guam between the Marshall Islands and the islands of the Federated States of Micronesia, just north of the equator, then take a convoluted path to the northwest, usually headed for the bigger land masses of the Philippines, Japan, or mainland Asia itself.

The Micronesian islands are of no consequence to these giant storm formations as they trundle across the Pacific. It makes no difference what little island is in their path; they just storm along, taking twists and turns as a matter of course. The Naval Typhoon Center on Guam is usually able to give several days' advance warning. This gives people time to run to the store for supplies, board up their windows, and tie down everything outside their houses. Airlines fly their aircraft out to a safer place, and all ships move away from the area as fast as they can.

When landfall of a storm is imminent, everything on the island closes down, and everyone goes home to wait out the storm. With gusting winds often reaching well over one hundred miles per hour, anything not bolted or tied down is hurled through the air and can be lethal to anyone out in the storm. Sheets of corrugated metal fly through the air like ballistic missiles. Coconuts become cannon balls.

The best way to pass the time is to throw a party at home and wait out the storm. In 1991, our last year on Guam, we were visited by Typhoon Yuri. The Typhoon Center gave plenty of warning, but the storm moved slowly toward the island, and the waiting became tedious. Finally, at nightfall, the siren winds started whistling around our house. During the night, we opened our window a crack to get fresh air, but the noise kept us awake. It sounded as if a never-ending freight train was passing just outside our window.

The storm is not the worst part of the drama. The aftermath is what brings mass depression to the people. Electricity is out because power lines are down, and this cuts off the pumps that distribute potable water to all the towns. Often, the government must send water-tank trucks into areas where no water is available.

But eventually things return to normal, and the island is a happy, sunny place to be again. Farmers whose banana and papaya trees were all destroyed plant the seedlings again and know that they will bear fruit in a matter of months. Trees that were felled by the storm are cut up for firewood, and sheet metal is removed from high in the trees.

During AWR's first ten years on Guam, no typhoon or earthquake was able to keep the station off the air for long. The worst damage occurred when Typhoon Omar made a surprise attack on October 28, 1992. Winds gusting to one hundred miles per hour forced water through the station's cooling system and into several transformers, which caused them to arc and burn out. Even with that, the station was only off the air for three weeks.

Then on August 8, 1993, the worst earthquake in eighty-four years, measuring 8.2 on the Richter scale, hit the island. The west end of the AWR-Asia building sustained moderate structural damage. Large cracks appeared in the wall, big chunks of plaster fell off, and some of the concrete blocks were broken. The tape library was a shambles, as thousands of cassette tapes were jiggled off their shelves and smashed to the floor. But even with such an upheaval, the station only lost seven hours of air time.

Being out in the middle of the ocean, as we were, helped us to understand a bit of what it must be like to be a sailor. We received many letters from people on ships. With little to do, they scan their shortwave radios for sounds to connect them with people on land.

A Korean sailor heard the Korean broadcast onboard ship. He wrote: "I have no religion but soothe my loneliness and the tediousness by listening to your broadcast. I'll be going home in five months. Your programs will help me make it through." ⊕

Another sailor, a Seventh-day Adventist from the Philippines, was the radio operator on a large container ship. He asked the captain for permission to put the AWR programs over the ship's public-address system. He was able to help his fellow sailors ease their loneliness by hearing their own languages—Ilocano, Tagalog, and Cebuano—over the radio. Soon ten crewmembers were joining him for regular Bible class, with the blessing of the captain, who liked the cooperative attitude his men were displaying since they started listening to the programs. Two ship's officers and several of the sailors planned to be baptized at the end of their tour of duty. ⊕

"Remember Zhi Ming?" Andrea asked one day, eyes dancing. "He chose the sailor's life. On board a cargo ship plying the South Pacific, he found time in his off-duty hours to listen to shortwave. He listened to the Mandarin broadcasts from Guam and became convicted that he should follow the Lord.

"Since he made that decision, he has found many ways to share his faith," she continued. "A fellow sailor, Chong, who was leading a totally dissipated life, joined Zhi one day to listen to a tape of hymns recorded off the radio program. The hymn that caught Chong's ear was 'The Ninety and Nine.'

" 'Could God love me like that?' he asked Zhi on impulse.

" 'Yes, of course He could,' replied his friend. Soon they were studying the Bible together. Chong became a changed man and started sending money home to his parents instead of spending it on cigarettes, beer, and other things.

" 'If God can make a change like that in my son,' said Chong's father, 'then we want to believe in Him too.' They soon became believers."

On another occasion, Zhi decided to go fishing while the ship was in port. "Nobody ever catches fish in this port," yelled his fellow sailors from the ship when they saw him with his fishing pole. "If you catch even one fish, I'll go to church with you," one man said.

Zhi smiled, threw his line into the water, and prayed, "Lord, help me catch some fish so that man will have to go to church with me." No sooner had he prayed than his line jerked, and he pulled in a very big fish! Then he caught two more. The other sailor not only went to church with Zhi but started studying the Bible with him too.

Once Zhi's ship was sheltered in a Philippine port because of a typhoon. In the dark of night, thieves boarded the ship and held up the sailors in their rooms with knives and guns.

"Give me your money," said one thief to Zhi.

"I will not only give you my money, but I will also give you my most precious possession. It will be your greatest treasure," he said as he held up his Mandarin-English Bible.

"Are you a Christian?" asked the startled thief.

"Yes," said Zhi. The thief took the Bible, but only half of the money. The other sailors lost all their money.

Some months later, while on shore leave in the same port, Zhi went to a restaurant for supper. While he was eating, the waitress brought him two more dishes of food. Surprised, Zhi protested, "I didn't order these."

"That man ordered them for you," the waitress said as she pointed to a man sitting near the window. Zhi looked over at a smiling man, whom he immediately recognized as the one who had taken his Bible many months before.

The man then stood up and walked over to where Zhi was seated. "Good to see you again." He was obviously anxious to start up a conversation. "I want to tell you about that money I stole from you," he began. "I took it that day and gave it to an orphanage that needed it. Do you have time to come visit the orphanage?" he asked.

"Yes, that would be great," replied Zhi. Together they walked several blocks to the orphanage, where a priest confirmed the gift of money and gave them a tour of the children's home.

"I always wanted to help an orphanage," wrote Zhi in his letter to us. "This was the way the Lord helped me do it."

Some months later, another letter came from Zhi, whom we affectionately called "our high-seas missionary."

"We have eight new sailors on our ship, all Vietnamese," he reported. "Please send me some spiritual material to give to them." Our traveling missionary was still at work.

Chapter 17

Island Farewell

"There's a Japanese gentleman here to see you. He says he talked with you by telephone about . . . buying the station?" Sharon Guth, my secretary, turned her matter-of-fact statement into a question.

"Oh yes, please ask him to come in," I said, then whispered, "and I'll tell you all about it after he leaves."

Mr. Takayama looked the typical prosperous executive. I was amused, but also a little disturbed, at his suggestion by phone that he was interested in buying our radio station. I sat down uneasily as we made ourselves comfortable around the big rectangular table in our "Japanese Room." We had decorated each room of the station with artwork from one of the countries to which we broadcast, and I was glad at that moment that this room had a large mural of cherry blossoms on silk and another picture with a Japanese motif.

"Our corporation purchased three hundred acres of land surrounding your radio station," he began confidently. "On this site we are planning to build four large hotels, hundreds of condominiums, an eighteen-hole golf course, a water park, and a shopping center." His words tumbled forth with lightning speed. I gripped the arms of my chair to control my panic.

"Because of these plans, we would like to offer you some land closer to the highway. Your old building here would become our clubhouse for the golf course!" His company had obviously planned thoroughly. He leaned back to see how I would accept his gracious offer.

"In principal, I have no objection to the plan at this time." I was feeling my way as I answered slowly. "However, you must first know that I do not personally own the station, and I do not have the final say. I believe we will need to schedule some more meetings to discuss your plan."

Thus began several months of negotiations that would keep us in a tizzy. We had difficulty comprehending that a move would be necessary but, at the same time, the company was offering a completely new physical plant and many additional dollars for our endowment fund to help us with long-term operational costs. We could also imagine that our radio station would have difficulty operating in the middle of a golf course if we refused to move. We prayed daily for the Lord's help with this new challenge.

AWR board chair Ken Mittleider came to lead out in our discussions with the corporation. Such a move, he cautioned, would be very costly and would take several years to accomplish. It would have to be done totally at the corporation's expense, without any loss of air time to AWR, and with considerable recompense for the inconvenience caused by the disruption of its operation.

It was difficult to concentrate on our work when such an upheaval was in the offing. The corporation's revelation was badly timed for us because an American Adventist businessman had just told us he would provide funds for a third transmitter. This was terrific news, but the possibility of a move forced us to put all other activities on hold.

In the end, the dilemma was resolved for us by financial conditions in Japan. Mr. Takayama's company fell into financial difficulties and had to abandon its development project on Guam.

At the same time, we were having great success with the two transmitters we had on the air. Results continued to pour in from all around Asia. Since we had gone on the air three years before, nearly forty thousand letters had come in from 114 countries.

One day in February 1989, in the Indonesian village of Sarireijo, a forty-seven-year-old farmer named Eliyas Subakir turned on his radio and heard AWR-Asia's broadcast in Indonesian. He was amazed to hear the Sabbath doctrine and thought it altogether strange.

Upon further Bible study, however, he became convinced of the truth in the message. He shared his new faith as he continued to listen to the radio. Soon he was leader of a group studying this new truth.

On May 5, 1991, Eliyas and his wife joined thirty-two others who were ready for baptism. Later in the year, their group grew by twenty-four. Most of the group were former Hindus and Muslims. The new church in Sarireijo is in the heart of one of the Adventist global mission areas, where there were no believers before. ⊕

A woman in China wrote: "I am a preacher in a large church. I think it would help me if I took your Bible-correspondence course." We sent the lessons, but the Bible school got no reply for many months.

Then one day, a large envelope arrived. It contained the answer sheets to all the lessons she had received! Her enclosed letter also had many questions. These were answered one by one. She felt compelled to share the good news of her new faith. Today, she hosts a new Seventh-day Adventist church in her home. ⊕

A young man in China listened to the broadcasts and recognized them as being from the church he had belonged to as a child. His heart was warmed, and he decided he must be baptized. He took the train, a two-hour trip, to Shanghai to find the Adventist pastor. He was informed that no baptism was planned any time soon in Shanghai.

"Maybe if you go to Beijing, they will have a baptism," suggested the pastor. The young man spent many hours on the train again, this time traveling to Beijing. But there, again, he was informed that no baptism would be held very soon.

"But I can give you the address of a pastor who can baptize you at any time," said the elderly pastor.

Again the young man was on the train. When he arrived, he told his story to the elderly gentleman who came to the door. He was questioned about his beliefs, and he showed the pastor his letter of referral from the pastor in Shanghai. The man then took him to his backyard, uncovered a tub that was dug into the ground, and baptized him.

This young man later left China to study in the United States. He then returned to Asia to help AWR make programs for his people. ⊕

The pastor of a church in China decided to attend an Adventist church service to find out why one of his strongest church members had become an Adventist. As he spoke with the Adventist pastor after church services, some new spiritual insights seemed to open before him. Convicted by the Spirit, he realized he, too, must follow this new religion that made such sense for his life.

He returned to his home church and started preaching what he had learned. Soon many members of his church became Adventist believers. The numbers continued to grow until six congregations were asking to become Seventh-day Adventist churches.

He wrote to the pastor who had originally introduced him to this new religion, asking for another worker to come help him lead the six congregations.

"We have a preacher for you," came back the reply. "He comes to us, too, every Saturday morning. We listen to him and are greatly blessed. You can hear him, too, if you listen to Adventist World Radio. We have dozens of big and little churches who have no pastor except the

Voice of Hope!" ⊕

As we thought of the contribution our station on Guam was making to the world church, we were extremely grateful for all those church members who had sacrificed financially to make it happen. Every day, checks large and small came in to support the radio ministry. More than the money, members' warm wishes and assurance of prayers encouraged us greatly.

"The Lord has blessed us, and we want to pass it on: here is a check for $250," wrote one retired couple. ⊕

A woman wrote, "I am interested in doing something in my husband's memory." Another woman declared, "I am 87 years old today, so I am sending you a birthday present to celebrate!" ⊕

"At long last our home has sold and we have collected the payoff. Accordingly, we have sent $10,000 for AWR," wrote one couple. ⊕

"I saved a little bit from my living expenses, and I would be glad if this could be used for the radio station," wrote one woman. She enclosed $500, enough to keep AWR on the air for four hours. ⊕

"I don't know how often I can send a gift. I'll send what I can," wrote one man. "I've always prayed for you, and I am thrilled about the mail you receive." ⊕

With such support, we were given wings to think bigger, to plan bigger. There was so much of the world to conquer for Christ. We still had burdens for those vast areas where we had no programs to share. One such country, the world's largest, had much of its territory in our coverage area. Russia sprawled over Europe and Asia.

Late in 1991, I was asked to visit Russia with Walter Scragg to see if we could assist in developing a new media center in Tula, Russia's first private media center.

Peter Kulakov, the media center director, was happy to meet two broadcasters who were anxious to see what had been accomplished. We found workers climbing all over the shell of a building. We advised Peter on how to best situate the radio studios and asked how soon programs in the Russian language would be ready for broadcast.

"As soon as our studio is operational, we will send you programs," he promised. Some months later, we were finally able to include Russian in our schedule from Guam.

Another potential audience whom we had always on our mind was right on Guam. The people of this hospitable island who had welcomed us as neighbors and friends could not hear our shortwave broadcasts. Our people on the island kept asking when we would have a station for them to hear.

I put the idea of a local station project on the back shelf of my mind because of other projects. Top priority was the church building for Agat. Then the Japanese corporation brought up its proposal of moving our station. We still had a third transmitter project on a front burner, and our fundraising program was advancing for that.

A main reason for establishing AWR-Asia was to broadcast to the main countries making up the territory of the church's Asia Pacific Division. One area that would not benefit from the shortwave broadcasts was the islands where the station was located.

Through my experience with AWR, I had developed a definite philosophy about new church-owned radio stations. I was embarrassed when I heard of cases in which the government opened the path for the church to establish local radio stations, but the opportunities were lost either because of lack of interest or a lack of confidence to take on the challenge. I determined that, as a representative of the church, I would do everything I could personally to see that an Adventist radio station was established whenever opportunity knocked. I would never be a silent voice when the Lord said, "Here is a way you can let My voice be heard; let this station be a lighthouse for Me in this community." I could not let the Lord down that way.

In cooperation with AWR, the Guam-Micronesia Mission planned to one day have a station on Guam, and I would encourage them however I could. They looked to AWR to help with the project. A three-kilowatt transmitter would reach all of Guam and Rota, the first Marianas island to the north, and later we could place a translator on Saipan, the second most populous island in Micronesia, about one hundred miles north.

Guam had my sympathy in this project for another reason. Already the island had seven AM and FM radio stations, all with rock or pop formats. We wanted to give the people something more beautiful. A good music station would add tremendously to the quality of life on the islands, and the religious programs would uplift the spirits of everyone who heard them.

With good financial support from two AWR board members who sympathized with the islanders, we began the project. Within a few months, we had financial commitments that would make the station possible. The hardware was assured.

But the software, the programming, would be the difficult aspect. On shortwave we could play programs in a different language each hour. But on local radio, many hours in the same language are needed. This is a completely different type of programming than what we were providing by shortwave to Asia.

AWR resources and personnel were stretched to care for the construction of the local station. The shortwave program kept us fully occupied, so we and other staff members donated many personal hours to getting the FM station running.

In 1991, JOY 92, KSDA-FM, took to the air over Guam and was immediately appreciated by people who had hoped for something different to listen to. The lieutenant governor called to say JOY 92 was his favorite station. Senators from the legislature claimed it was their favorite station too.

"Life is so good on Guam, and we've been able to accomplish so much, do you think we could ever live anywhere else?" I put a hard question to my wife one day.

"It would be hard to beat Guam, wouldn't it?" she asked in return. "But I did enjoy Europe very much, didn't you?"

"Yes, I really did," I admitted. "I guess if we are ever given the choice of where to go next, I would say Europe too." We had no idea why we were having such a conversation, but it all became clear some weeks later.

"I've just been on the telephone with Walter Scragg," I announced to Andrea. Walter had been called back to head the church's radio ministry after serving a term as leader of the South Pacific Division. Now as president of AWR, he was inviting us to a new post as AWR system directors located at AWR's headquarters in Europe.

Andrea could hardly believe her ears. We looked at each other and said in unison, "Yes." Life was good on Guam, and we enjoyed our work there, but after seven years, maybe it was time for a change. The day-to-day pressure of administrative responsibilities had begun to wear me down, so the thought of starting over with different responsibilities seemed like a breath of fresh air.

Soon we were filling boxes with household goods, readying for our move to a new challenge, a new job. But we both loved AWR so much we were glad it was a move to help in this radio ministry. Andrea was to be public-relations and development director, and I would be program director for the AWR system. We felt we were on the verge of an exciting new chapter in our lives.

Chapter 18

Europe Landing

"And you say it has nice big windows, a balcony, and it's only a ten-minute walk to the office?" Andrea asked, following a string of other questions. She could hardly believe I could find such an apartment after all we had heard about the difficulties newcomers encountered when looking for a place to live in crowded Germany.

"In the worst-case scenario, you may have to purchase and install your own kitchen—sink, cupboards, and all," Walter Scragg had warned us. But just like Andrea, he was pleasantly surprised when we reported that we had found a furnished apartment conveniently located on the main road, Heidelberger Landstrasse, between the city of Darmstadt and its suburb, Eberstadt, just twenty-five kilometers from the large city of Frankfurt am Main.

Andrea was still amazed when, on our first day in Germany, we climbed the stairs and entered the nicely appointed four-room apartment. She walked from room to room to see and touch every counter, chair, and cupboard.

"I think it will be good for us to have an apartment; it will give us more time to work for AWR," she added as she started unpacking our clothes and placing them neatly in the bedroom chest of drawers.

"It will be good, too, not to have a yard to maintain," I

added, thinking of the extensive travel listed on my first year's itinerary. We were grateful the Frankfurt Airport was only a twenty-five-minute drive away on the autobahn or a half-hour trip by Airporter bus. Yes, we decided, this would fit very well into our busy lifestyle.

We liked our landlady, Frau Scharf, as much as our apartment. She was a widow and welcomed us as if we were family and then proceeded to spoil us as if we were her very own. She developed the habit of cooking a lovely fruit torte for us whenever we returned from a long trip. She often asked us to join her and friends for supper on her back terrace overlooking a private garden that displayed a wide variety of flowering plants and fruit trees. Set in the middle of the garden was a small fountain that gurgled down into a little pond filled with water lilies and goldfish.

AWR had only recently made the decision to move its European headquarters to Darmstadt to take advantage of its central location for travel and communication. Greg Hodgson was general manager for the region, Claudius Dedio chief engineer, Pino Cirillo assistant engineer, and our assignment, not so easy a task, was to work partially with them in the European region and partly for the worldwide AWR system. None of us were sure how this arrangement would work, but we were happy to be together, and this made us willing to try anything.

It would take us some time to grasp everything that had happened to AWR in Europe in recent months. The continent was in a turmoil of change, caused mostly by the collapse of Communism in eastern Europe. The two Germanies were reunited, and numerous countries of the former Soviet bloc had either achieved independence or were working toward it.

AWR suddenly had the opportunity to begin broadcasting in the former Communist countries. The Adventist Church had been foremost in advancing its activities as the doors of freedom opened. It had been first to establish a theological seminary in Russia, first to build and open a printing house, and first to go nationwide with radio and television programs. These efforts had positioned the church well for the requests that AWR would make of the powerful Radio Moscow stations.

"Almost unbelievable," is how Walter Scragg described an offer from government officials in Russia. In response to inquiries about broadcast possibilities, AWR was offered the opportunity to lease a transmitter in Siberia for broadcasts into the Indian subcontinent.

A contract between the Voice of Hope Media Center in Tula, Russia, and the cooperative managing the powerful propaganda shortwave station in Novosibirsk, Siberia, would take AWR programming to the airwaves twenty-three hours daily.

As it turned out, the facility at Novosibirsk was probably the very station that had been used to jam AWR Russian and Ukrainian programs from Radio Trans Europe nearly twenty years earlier. It was also one of the world's largest radio stations, boasting thirty transmitters, with 120 curtain antennas spread over two thousand acres.

"We live in a day of miracles," said AWR board chair Kenneth Mittleider at the inauguration of the Novosibirsk broadcasts on March 1, 1992.

"Two years ago, such a possibility would have been laughed to scorn, yet today it is a reality."

"It was bitingly cold in Siberia," Walter said about the inauguration of the broadcasts. "It was forty below zero, centigrade and Fahrenheit [forty below is the one place where the two temperature scales read the same], but we were warm inside because of this great occasion." He described the emotional experience as AWR became the first Western broadcaster to go on the air on the former Communist propaganda stations. A woman in the city church stood up during the service to tell how for fifty years she had prayed that those mighty transmitters she could see in the distance would one day be used by the Lord for His purposes. There were very few dry eyes as she gave her testimony.

Soon contracts were signed for use of other Russian stations at Moscow, Ekaterinburg (Sverdlovsk), and Samara (Kujbysev). It started a boon for Western broadcasters and was a new source of income for the stations that now had no propaganda programs to broadcast. Soon other Western organizations followed AWR's example and found unused transmitters around Russia from which to serve areas of the world that had been difficult for them to reach before.

With the advent of programming from Russia, the end of an era came for AWR at the Radio Trans Europe site in Portugal. On Sunday, June 28, 1992, AWR broadcast its final programs from Portugal. For twenty-one years, this station had been our major voice for a large part of the world. But because air time costs had climbed too high, programming had diminished to only seven-and-one-half hours per week. By designating fourteen of the hours we had just licensed on the Russian stations for the programs that used to come from Portugal, we could nearly double the amount of time given to these broadcasts. Our increased schedule brought us to the attention of the short-wave community as "a major international broadcaster."

While we were rejoicing with the new developments in Russia, stories of success continued to come from other parts of the world. Not just individuals, not just families and villages, but whole churches were responding to the gospel message they heard on AWR.

In a country in Indochina, pastors of three churches wrote to say that they used material from AWR sermons for their own sermons. One church group even used the AWR broadcast on Wednesday evenings for their prayer meeting.

The coordinator for an independent gospel ministry in Sri Lanka wrote: "We have ten branch churches, and we extend expertise and advice to six emerging churches. We have a number of home cells and other enterprises to take the love and redeeming power of our Lord to our neighbors. We also do a lot of counseling. In spite of civil strife here, people are drawing closer to our Lord. We would like to get transcripts of your broadcasts." ⊕

The national president of a church in Nigeria wrote: "We are members of a Bible-believing and practicing church. Having listened regularly to your radio broadcast, and having examined critically your doctrinal expositions, we have discovered that we have identical visions and beliefs. Consequently, we are requesting an affiliation with your ministry."

And members of another church in Nigeria said: "We wish to let you know that we are avid listeners to AWR. We listen to your evening programs and take it as our evening studies. Long live AWR!" ⊕

Back in Europe, we heard of a young man in the Polish military who had an incredible experience with AWR. He was in training as a paratrooper. His workweek consisted of repeated parachute jumps.

"One day we were up in the air for one of our regular practice jumps," he wrote. "As I jumped, something happened that had never happened before. The guy right behind me was not paying attention and jumped almost immediately after I did. As our parachutes opened up with the whip of the wind, they became incredibly entangled. We both were fighting desperately to untangle those parachutes while falling faster and faster toward the ground.

"All the time, one question was racing through my mind: Is this it, God; is this the end for me? What will happen to me when I die, God? What will happen?

"Finally, we don't know how, we were able to untangle ourselves enough to jettison the entangled chutes. We got rid of that extra baggage just in time to open our emergency chutes. We landed on the ground with a thud.

"But the question in my mind persisted: What would happen if I die?" He was incredibly shaken by the accident. That night, to try to forget the experience, he turned on the radio. The first program that he heard in Polish was the *Voice of Hope* on AWR. The topic: "What Happens When You Die?"

He couldn't believe his ears. Of course, he signed up for the Bible course that was offered. Later, he became a member of the Seventh-day Adventist church in his city. ⊕

Another exciting activity was taking place on the island of Cyprus, where the church was preparing a media center to produce programs for the Middle East. AWR had broadcast for many years in Arabic, but the programs were especially prepared for North Africa. Now we would begin a major effort for the Middle East itself.

With the election of Robert Folkenberg as president of the General Conference in 1990, Neal Wilson handed over the reigns of leadership. But he was determined to be of service to the church in his retirement years. One of his first thoughts was to help advance the radio ministry in the Middle East. As a young man, he had served his first overseas assignment in that region.

Church members were invited to contribute to a fund to build the Cyprus media center to honor Wilson for his years of service to the world church. The goal for the project was $25,000, but he was pleased to learn that over $35,000 was donated. Within a short time, with Bert Smit directing the Adventist Media Centre-Middle East, AWR was able to broadcast up to eight hours a day of Arabic programs into the Middle East.

Bert was desperate for program materials for the many hours of new programs in Arabic. He heard rumors that some old program tapes existed somewhere, produced many years before for the broadcasts from Portugal. They had disappeared during the war years in Beirut, Lebanon. Different attempts to find the tapes were fruitless.

While visiting in Beirut in early 1993, Bert and his colleague, Amir Ghali, were poking around an abandoned church office in the city. The building had been extensively damaged by mortars during the civil war. There, in an old cabinet in the corner of the building, they discovered the lost treasure. Providentially, dozens of tapes in Arabic, Farsi, and Turkish were there just waiting for someone to find them.

"There is a whole series on health, mostly still relevant today," said Bert. "The classical Arabic is of very good quality, and with minor editing, we can use all the tapes. It is a wonderful find, and we are grateful to God for it." ⊕

In Latin America, AWR made a different kind of find. There, the government had built a radio station to counteract rebel insurgents in Central America, but now the political situation had stabilized, and the station was classified as surplus equipment. David Gregory, manager at AWR headquarters in Alajuela, Costa Rica, heard of the pending sale and put in an early bid for the station.

The result was an incredibly small purchase price for very valuable property. With some renovation, the station would have five transmitters, with maximum power of fifty kilowatts each, to cover all of Central America and adjacent parts of North and South America. ⊕

In Africa, AWR was continuing its expensive purchase of air time on Radio Africa 1 in Gabon. The most powerful station on the continent, it provided an excellent signal into west and central Africa. Seven hours of programs a week were not much, but results were heartening.

Meanwhile, church leaders continued to search for a site to build our own station in Africa. Various sites had been investigated, beginning with Liberia. But the revolutionary war there in the early 1990s made us drop that option. By 1995, two sites were strong possibilities:

Namibia and São Tome. We hoped for a definite sign, a breakthrough that would make clear our path for Africa. Meanwhile, one hour a day on Radio Africa 1 was the best we could do. Listeners established three new churches: two in Benin and one in Togo. Correspondence with a group at Ifangny, Benin, began in 1990 when three members of another church wrote to AWR. These three people were cast out of their former congregation when they started sharing the message they heard on AWR. They formed their own group with four families and some friends and were meeting in a rented garage they had fixed up as a church. ⊕

From Azove, Benin, a letter written on behalf of seventy people said: "We have received the whole good news in its purity." The group was already active in witnessing by organizing a musical guitar group. They presented public concerts and gave short sermons for those who attended. ⊕

Eight family members in Notse, Togo, joined workers from the Adventist dental clinic in Glei to form another congregation. The new church was located seventeen kilometers from Lome, the country's capital. ⊕

At another village in Benin, an Adventist man was selling books house to house, searching for people who had graduated from the AWR Bible-correspondence course to see if they cared to purchase any religious literature.

"Yes," said the man who opened the first door, "we are Seventh-day Adventists! Come, let me show you something."

Together they went to a small church building, where the man pointed to the sign over the door, which read "Seventh-day Adventist Church." He turned the sign over—there was the name of another denomination. Nearly the whole church of seventy members had become Adventists!

Chapter 19

AWR's
Silver Lining

The early 1990s turned out to be banner years for AWR. Weekly hours on the air topped one thousand, the number of transmitters in use expanded to sixteen, and nearly fifty studios were busy producing programs in forty languages.

"There's an amazing thing about all this," Walter Scragg said to us one day when he and his wife, Elizabeth—a much appreciated member of the AWR team—were visiting us in Darmstadt. "Of all the major broadcasters, we probably have the smallest staff! We have become adept at doing much with very few resources." I had to admit it was true. Under Walter's leadership, we all developed a motto of "put the money into broadcast hours."

Programming, of course, was my full-time concern. My part-time responsibility for systemwide programming turned into a full-time vocation as AWR expanded and needs around the world increased. I was also responsible for training and audience research. With Walter's encouragement, we were able to establish a program resource office at Newbold College, England, to provide materials to help program producers create attractive programs. Ray Allen, a former student missionary at AWR, was asked to lead out there.

In 1993, we produced our first systemwide program schedule. Previously, each station had provided listeners with a regional schedule. Now we were starting to present ourselves as a unified worldwide entity. The result was new respect by the international broadcast community and a better identification of AWR's broadcasts by listeners around the world.

AWR developed region by region and was administered according to five geographical areas. AWR-Europe came first when programs began in 1971 from Portugal. AWR-Asia and Latin America began in the late 1970s under pioneers Adrian Peterson and Robert Folkenberg, respectively. Daniel Grisier led out in developing AWR-Africa in the 1980s, and AWR-Russia developed in the early 1990s under Peter Kulakov.

AWR had the courage in 1994 to convene a planning commission to study how its operation could be streamlined. An attempt to minimize duplication of effort was made by giving authority to a central AWR administration. Despite the challenge of reorganization, AWR was able to concentrate on its mission surprisingly well. A talented core group of managers and directors made things happen. I learned to appreciate and value the contribution of each of my colleagues.

"Take Greg Hodgson, for example," I said to Andrea one day as we were discussing our work. "His persistence in trying to get permission to build an AWR station in Italy is incredible." For the 1990 General Conference offering, church leaders had agreed to create a fund for building the station, and in 1995 the funds were still in an account waiting for the day when the new station could be built. Greg continued urging and encouraging the Italian government to pass legislation permitting such a station.

"Greg has been working to make it happen for five years now," I continued. "He's like a dog with a bone; he just won't give up."

"Well, don't forget, either, his role in getting us on Radio Slovakia," Andrea added. "Where would we be today without that station?"

Radio Slovakia came as a heaven-sent opportunity for AWR in 1994. In the city of Rimavska Sobota, meaning "Roman Sabbath," the Communists had built one of the most effective international stations in the world. It was known as Radio Prague during its propaganda days, but now, with the peaceful split of Czechoslovakia into the Czech Republic and the Republic of Slovakia, this incredible facility became a prime possibility for use by AWR.

As soon as possible after the great political changes in eastern Europe, Greg began probing officials in Prague and Bratislava about the possibility of purchasing air time on the Rimavska Sobota station. It had powerful 250-kilowatt transmitters connected to some of the world's largest and most effective antennas, known as 8 x 8 curtains. This gave it a tremendous broadcast range, reaching to Africa, Asia, Europe, the Americas, and the Middle East.

Repeated contacts with Slovakian Radio bore fruit in late 1993 when AWR was granted a license to broadcast over stations in the republic. AWR immediately contracted for full-time use of two transmitters.

The opening came none too soon. The Russian stations, which had bargain prices for air time when they first became available, drastically increased their prices year by year and by 1994 began to price themselves beyond AWR's budget. This eventually became a problem in Slovakia, also, when prices doubled within two years. Nevertheless, test broadcasts from Rimavska Sobota began on January 1, 1994, and regular broadcasts began on January 8, when AWR leaders joined state officials in Bratislava for an official opening ceremony.

The Radio Slovakia station expanded our capabilities

tremendously. For the Middle East, we now had a signal that boomed into the area like a local station.

"During the first quarter of 1995, a record three hundred letters arrived at the Bible School from listeners in the Middle East," reported Bert Smit. "We had letters from every country in the region, and a considerable number came from countries where the church has no established work," he added. His report was augmented by the Paris studio, where Arabic programs are produced for North Africa. A record mail count was reported there as well, and records at both places continued to topple as mail increased.

"In one country, two listeners write that they are studying the Bible course together," Bert continued. "In another country, a group of twenty listeners formed a *Voice of Hope* Radio Club. They listen to AWR together on a regular basis. But in other more restricted areas, new believers worship secretly around the radio."

Reports from Baghdad, Iraq, indicated that a group of AWR listeners augmented the church there when they started attending weekly worship services. Most first came to the church to ask for the books promised on the Arabic radio programs.

In Europe, letters from listeners increased significantly in response to the English and German programs and new programs in the Czech, Hungarian, and Bulgarian languages. AWR, thanks to Slovakia, was on new footing. Only having our own station would be better.

"Your programs are always welcome," wrote a listener in Italy. "Your programs have an open mentality that does not create barriers with nonbelievers." ⊕

"I can honestly say that of all the religious broadcasters, AWR is my favorite for a balanced message—without a lot of shouting evangelists—and you have nice music and interesting talks," offered a listener in England. ⊕

"For as bad a man as I am, your program is the most

uplifting that I have found in the dump of radio stations," a listener in the Czech Republic wrote. ⊕

Our withdrawal from the shortwave stations of the former Radio Moscow didn't decrease our listenership in Russia. Our media center, with initial financial help from the Voice of Prophecy in America and a long-term commitment by AWR, was able to obtain program time on national radio networks in all the countries of the Commonwealth of Independent States.

Most homes in the former Soviet Union had been wired for radio somewhat like a cable television system, and this created one of the world's largest radio audiences. AWR's *Voice of Hope* became the most popular religious radio program in Russia, according to a national audience research project that found we have four million listeners each week!

Listener response was so heavy that the media center had difficulty processing the mail. Not only the media center, but the post office as well.

"You have nearly four thousand letters here," the postmaster reported by telephone one day. "Would you please come and get them," he pleaded, "so we can sort the rest of the mail?"

In a short time, the media center could point to many new churches around Russia that had their beginning from *Voice of Hope* broadcasts. In the city of Tula itself, the radio broadcasts brought in a large part of the membership of the new Central Seventh-day Adventist Church.

The fifty-sixth General Conference session in Utrecht, Holland, in the summer of 1995, provided a fitting climax to five incredible years for AWR. Gathered there, to cover the historic event in twenty languages, were nearly fifty Adventist radio broadcasters.

To accommodate the radio corps, we constructed our own broadcast center. It was an impressive structure located at a crossroads of session activities. It housed seven recording studios, a master control room for sending out radio programs, a local radio station, and offices. These were all clustered about a large work area that contained computer editing desks and recording edit stations. The whole center had windows strategically located to provide viewing from outside, where streams of people passed by constantly.

Just outside the main-hall entrance, we positioned our rented satellite uplink dish. Day and night, programs went up to a satellite and from there to the transmitters in Slovakia, which broadcast them to Europe, Africa, Asia, and the Middle East. Programs for the Americas and Asia were sent through a combination telephone/satellite system to our stations in Costa Rica and Guam.

The session broadcasts were a huge success. Some people now attend Adventist churches because of these special broadcasts. Three young Muslim employees at the Jaarbeurs Convention Centre, where the session was held, signed up for Bible studies at the AWR exhibit booth.

"Well, we survived this big event," Andrea said when the session was over. "Now let's see how we do on the next one." She was referring to our move to Britain. In Germany, the deutsche mark had reached new highs against the U.S. dollar, and the cost of operating in Germany was becoming prohibitive. To economize, AWR decided to move offices to Britain.

It was the shortest move we have ever been asked to make. A quick flight of just over an hour from Frankfurt to London landed us in our new country. Ahead was the job of finding an office and a place to live and a search for personnel to help at the office. Newbold College, two years earlier, had supplied a small office for our program resource office, but there was no more room for other personnel there. We immediately began a search around the Binfield community, a search that would last several months.

Then we were on our way to the annual AWR meetings at the General Conference office near Washington, D.C. A change of presidents for AWR was to take place, we would inaugurate a new board of directors with new chair, Philip Follett, and we would have a once-every-five-years constituency meeting.

We were only a few days into our round of meetings when an urgent message came from England. A fire had broken out in the warehouse where our advance container was in storage! According to police, the entire warehouse was consumed by the fire, probably caused by a fault in the electrical wiring system.

"Oh no," Andrea and I both gasped, "all our office files were in that container." We stumbled through several days of shock, every minute trying to remember what actually had been shipped in that container, which was one of five. Most of our clothes were in the box, as were many boxes of office files, books, and materials. We weren't concerned about the clothes—we could tolerate a new wardrobe—but many of the files would be impossible to replicate. It was a serious loss, and it would take us some time to determine its extent. We later confirmed, we believe that many of the historical records used in writing this book may have gone up in the flames. We are thankful that we were able to nearly complete the manuscript and send it off to Pacific Press prior to our move!

Meanwhile, our concentrated meetings turned out to be remarkable. Remarkable because of the standing that the radio ministry had achieved in the church and remarkable because of new plans being laid. A special challenge was to find ways to cut $700,000 from the 1996 budget. Budget requests were that much over the projected income, but with the help of our new controller, Richard Green, we were able to present a balanced budget to the board.

Gordon Retzer, new AWR president, speaking at the

constituency meeting on October 4, pledged to operate AWR on as economical a basis as possible, to add many new languages, and to use new technology to carry out our mission.

"To operate economically," he said, "we will concentrate on keeping overhead expenses lean so that we can fund an expanding broadcast schedule." He focused on a centralized concept of administration as AWR proceeded with restructuring to decrease administrative activities at radio stations.

"AWR accepts the challenge of increasing languages so we can reach a larger share of the world's population," Retzer stated. "This will involve adding over fifty new languages to our schedule by the year 2005." He referred to the Adventist Church's plan called "Global Mission," which targets areas of the world where the church has little or no presence.

I was excited to hear him commit AWR to the use of new technologies. "We are interested in utilizing any new technology that will help AWR accomplish its mission," he said. He specifically mentioned use of satellite for distribution of programming.

During these same meetings, the AWR board of directors approved a plan to create a fellowship for listeners. This is an organization that listeners who identify with the message we broadcast can join. We will periodically send information and words of encouragement to fellowship members to help them feel part of the AWR family until a more personal contact can be made.

We also received unbelievable news from Italy. After nearly six years of petitioning by AWR, the Italian government put into effect, on October 4, 1995, a new law permitting licensing of shortwave broadcast stations on Italian soil.

Until this law went into effect, Italy had no regulations for international broadcasting. AWR's little shortwave station in Forli, however, had been operating since 1984, with the hope that official recognition would someday be granted.

We had voted at our meeting to cease our pursuit of a license in Italy and close the Forli station, but God turned that decision around! We rejoiced again in His leading. The story of AWR is a story of God's leading step by step. Once again, we had solid evidence of that.

"Every day, we exult in how God is leading, where He is pointing," Walter wrote as he summed up AWR's situation. "We read the letters, check off the new languages, and thank God for the support and prayers!"

We decided that our lives are mere silhouettes on a large canvas backdrop that God is painting full of the color of faces from every corner of earth. The faces of those attracted to Him through the miracle radio waves that leap national boundaries, overcome prejudices, and penetrate walls, both real and imaginary, with the message of His love.